Collins

AQA GCSE 9-1
Geography
Revision Guide

Paul Berry, Brendan Conway, Janet Hutson, Dan Major, Robert Morris and Iain Palôt

About this Revision & Practice book

Revise

These pages provide a recap of everything you need to know for each topic. All key words are defined in the glossary.

You should read through all the information before taking the Quick Test at the end. This will test whether you can recall the key facts.

Practise

These topic-based questions appear shortly after the revision pages for each topic and will test whether you have understood the topic. If you get any of the questions wrong, make sure you read the correct answer carefully.

Review

These topic-based questions appear later in the book, allowing you to revisit the topic and test how well you have remembered the information. If you get any of the questions wrong, make sure you read the correct answer carefully.

Mix it Up

These pages feature a mix of questions for all the different topics, just like you would get in an exam.

Test Yourself on the Go

Visit our website at **collins.co.uk/collinsGCSErevision** and print off a set of flashcards. These pocket-sized cards feature questions and answers so that you can test yourself on all the key facts anytime and anywhere. You will also find lots more information about the advantages of spaced practice and how to plan for it.

Workbook

This section features even more topic-based questions as well as practice exam papers, providing two further practice opportunities for each topic to guarantee the best results.

ebook

To access the ebook revision guide, visit **collins.co.uk/ebooks** and follow the step-by-step instructions.

QR Codes

Found throughout the book, the QR codes can be scanned on your smartphone for extra practice and explanations.

A QR code in the Revise section links to a Quick Recall Quiz on that topic. A QR code in the Workbook section links to a video working through the solution to one of the questions on that topic.

Contents

	Revise	Practise	Review

Geographical Skills

Geographical Skills — p. 4

The Challenge of Natural Hazards

Tectonic Hazards — p. 8 — p. 22 — p. 38
Tropical Storms — p. 14 — p. 23 — p. 38
Extreme Weather in the UK — p. 18 — p. 24 — p. 40
Climate Change — p. 20 — p. 25 — p. 40

The Living World

Ecosystems and Balance — p. 26 — p. 41 — p. 62
Ecosystems and Global Atmospheric Circulation — p. 28 — p. 41 — p. 62
Rainforests and Hot Deserts – Characteristics and Adaptations — p. 30 — p. 42 — p. 62
Opportunities, Threats and Management Strategies in the Amazon — p. 32 — p. 42 — p. 63
Opportunities, Threats and Management Strategies in Hot Deserts — p. 34 — p. 43 — p. 63
Polar and Tundra Environments — p. 36 — p. 43 — p. 63

Physical Landscapes in the UK

Coasts 1: Processes — p. 44 — p. 64 — p. 86
Coasts 2: Landforms — p. 46 — p. 64 — p. 86
Coasts 3: Management — p. 48 — p. 64 — p. 86
Rivers 1: Processes — p. 50 — p. 65 — p. 87
Rivers 2: Landforms — p. 52 — p. 65 — p. 87
Rivers 3: Flooding and Management — p. 54 — p. 66 — p. 87
Glaciation 1: Processes — p. 56 — p. 67 — p. 88
Glaciation 2: Landscape — p. 58 — p. 67 — p. 88
Glaciation 3: Land Use and Issues — p. 60 — p. 67 — p. 88

Urban Issues and Challenges

Urbanisation — p. 68 — p. 89 — p. 112
Urban Issues and Challenges — p. 70 — p. 89 — p. 112

The Changing Economic World

Measuring Development and Quality of Life — p. 76 — p. 90 — p. 113
The Development Gap — p. 78 — p. 90 — p. 113
Changing Economic World Case Study – Vietnam — p. 80 — p. 91 — p. 114
UK Economic Change — p. 82 — p. 91 — p. 114
UK Economic Development — p. 84 — p. 91 — p. 114

The Challenge of Resource Management

Overview of Resources – UK — p. 92 — p. 115 — p. 118
Food — p. 94 — p. 115 — p. 118
Water — p. 98 — p. 116 — p. 118
Energy — p. 102 — p. 116 — p. 119

Geographical Applications

Fieldwork — p. 108 — p. 117 — p. 120
Issue Evaluation — p. 110

Mixed Questions — p. 121
Answers — p. 128
Glossary and Index — p. 142

Geographical Skills

You must be able to:

- Demonstrate a range of geographical skills including cartographic, graphical, numerical and statistical.

Cartographic Skills

Latitude and Longitude

- All points on Earth can be located precisely with coordinates using **latitude** and **longitude**.
- Latitude lines provide a measure of how far places are north or south of the Equator. Latitude lines are all parallel to the Equator.
- Longitude lines provide a measure of how far places are east or west from the Prime Meridian (sometimes called the Greenwich Meridian). Longitude lines run between the North Pole and South Pole (meeting at the Poles) at right angles to the latitude lines.

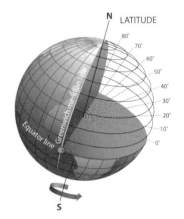

Atlas Maps

- Categories of map include:
  - General: physical (showing **relief**, drainage); human (showing population distribution, population movement, transport networks, settlement layout); and **political** (showing borders).
  - Thematic: weather; climate; ecosystems; demographic (population); development; economic (industry).
  - Environmental issues: climate change; pollution; desertification; deforestation.

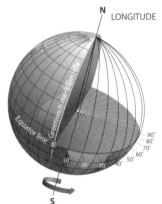

Patterns and Distributions

- You will need to be able to recognise:
  - Patterns: where are things concentrated in **dense** clusters or more **sparse**?
  - Patterns found in: land use; populations; settlements; communication networks (transport and telecommunications); earthquakes; tropical storms.
  - Patterns that have recognisable shapes, e.g. **nucleated**, **linear**, **dispersed** or **radial**.
  - Distributions: describe where things are in a broader sense, using compass points and terms such as central, coastal and **peripheral**.

Ordnance Survey (OS) Maps

- You should be able to locate places using grid lines: give the number of the horizontal **easting line** first, then the vertical **northing line**. A rule may be useful, such as 'along the corridor and up the stairs'.

- Use four-figure grid references to locate whole grid squares – this is best for larger areas.
- Use six-figure grid references to locate specific points.

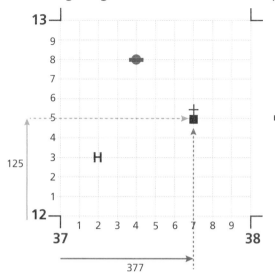

The six-figure grid reference for the church is:

377125

Ordnance Survey mapping

Scale

- All maps use **scale** to provide a standardised way of representing the real world on a page or device screen. Map scales often use a : (colon) format, which means 'represents' or 'equal to in reality'. Common OS map scales are 1:25000 and 1:50000.
- Use the scale to check your technique for measuring straight- and curved-line distances on maps.

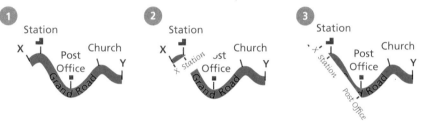

Compass Directions

- Learn the points of the eight-point compass and practise using them to describe locations, e.g. 'the camera is pointing south-west'; 'the river flows in an easterly direction'.

Isolines as Used on Relief and Meteorological Maps

- Relief is the shape or topography of the land, shown by isolines called **contours** or by layer colouring.
- Height is usually measured in metres above sea level.
- Points of exact height are shown using **spot heights** or **triangulation pillars**.
- Contours are lines along which the height is the same.
- Steep slopes are shown by closely spaced contours; gentle slopes are shown by contours with larger spaces between them.
- Landscape shapes include **valleys**, **ridges**, **spurs** and **plateaus**.
- Descriptions can also refer to their shape, size, elevation and compass direction (**orientation**).
- Cross-sections are like slices through a landscape, drawn using contours. These can be drawn by hand or with **GIS** (**geographical information system**) elevation profiles.

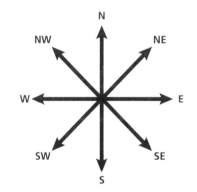

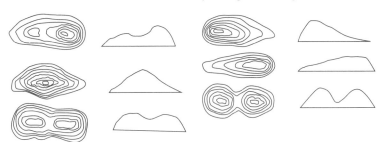

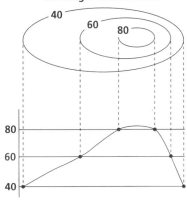

Imagine walking along a straight line from the west side of the hill to the east. You can draw a transect line to show your route. On the transect, put a small dot each time a contour is crossed and label it with its height. Use the line to plot the heights on graph paper or a chart. Then 'join the dots' to show the relief of the hill.

- Helpful terms to describe hills or valleys include **symmetrical** (similar gradient on both sides) or **asymmetrical** (steeper on one side than the other).
- Relief has a significant impact on human activity, e.g. land use and communications.
- **Isolines** are often used to show weather data such as pressure (**isobars**) and rainfall (**isohyets**).
- They can also be an effective way to map data such as pedestrian counts or pollutants.

Map Symbols

- Always refer to the symbology of maps, using the key provided.
- Practise interpreting by making links between symbols or their absence. For example, on an OS map, a lack of human activity on land beside a river (where there are large gaps between contours) could be due to a desire to avoid flooding.

Drainage: Water on the Land

- River flows or an absence of rivers tell us about the drainage of water from a landscape.
- Drainage density is measured by the total length of river channels in an area, e.g. km per km², and is usually high in areas with **impermeable** rocks and low in areas with **permeable** rocks.
- On some river channels, **hard** or **soft river engineering** is used as a form of **flood management**.
- Drainage includes lakes, waterfalls, underground rivers (e.g. in limestone areas) and artificial water features such as reservoirs.
- Terms used to describe drainage patterns include dendritic, rectangular, parallel, trellised, deranged and radial:

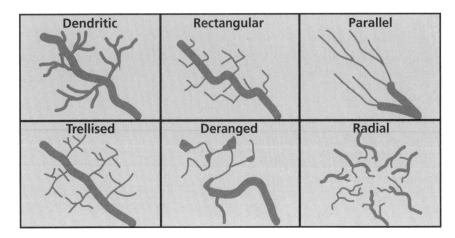

> ### Key Point
>
> You will use your geographical skills when answering questions about all the topics you have studied, and there are lots of opportunities to practise these skills throughout this book.

> ### Key Point
>
> Fundamental geographical skills include the ability to use and understand maps, graphs and charts, and being able to employ numerical data and statistical techniques effectively.

Graphical Skills

- You should be able to suggest appropriate forms of graphical representation; interpret and extract information from, and add to, the following:
  - scatter graphs; line charts; pie charts; bar charts; **histograms** with equal class intervals; particular types of charts such as population pyramids, divided and cumulative bar and line charts

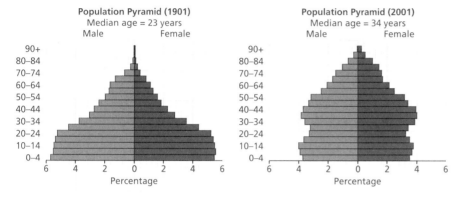

  - pictograms; proportional symbols; **choropleth maps**; isolines; dot maps; desire lines; and flow-lines
  - aerial imagery; satellite imagery; false colour images.
- Fieldwork can often be enhanced by photographs and field sketches, which are really 'diagrams' of a place. Good annotations can add value to graphics.
- Sketch maps are roughly drawn maps that show basic details of an area, and can be labelled or annotated to identify key features.

Numerical Skills

- Number, area and scales; **quantitative** relationships between units.
- Data collection: accuracy; sample size and procedures; control groups; reliability.
- Proportion and ratio; magnitude; **frequency**.

Statistical Skills

- Measures of central tendency: **median**; **mean**; **mode** and **modal class**.
- Measures of spread and cumulative frequency: range; **quartiles** and interquartile range; dispersion graphs.
- Percentages: percentage increase or decrease; **percentiles**.
- Relationships in bivariate data: trend lines through scatter plots; lines of best fit; positive and negative correlation; strength of correlation.
- Predictions and trends: **interpolation**; **extrapolation**.
- Limitations and weaknesses in selective statistical presentation of data.
- Geospatial (geolocated or georeferenced) data presented in a GIS framework. GIS can also be used to analyse spatial data.

Tectonic Hazards 1

You must be able to:

- Understand that natural hazards are the result of physical processes
- Understand how tectonic hazards pose a risk to people and property
- Identify ways that effects from tectonic hazards can be reduced.

Natural Hazards and Their Impacts

- A natural hazard is an extreme event that causes loss of life, severe damage to property or severe disruption to human activities.
- Natural hazards can have a social, economic and environmental impact on an area, and they are more severe in lower income countries (LICs) because there is:
  - more poor quality housing and poor healthcare
  - poor infrastructure so it is harder to reach affected people
  - less money to protect people (e.g. with earthquake-proof buildings) and less money for responses (e.g. providing food).

Earthquakes and Volcanoes

- The Earth's crust is divided into seven large plates and 12 smaller plates that sit on the mantle below. They move very slowly in different directions due to heat convection currents in the mantle:
  - Magma below the surface is heated by the core and rises to the crust, where it cools and sinks, to be reheated once more.
  - Friction caused by the rising currents means that the plates are pushed forward or dragged back (plate tectonics).
  - There are two types of crustal plate: oceanic plates (tend to be denser but thin) and continental plates (less dense and thicker).
  - Two (or more) plates meet at a plate boundary or **plate margin**.
- There are three main types of plate boundary:
  - **Constructive margins** (or divergent boundaries) are where plates pull apart, allowing magma to reach the surface and erupt through fissures and faults, creating volcanoes. Earthquake activity also occurs at constructive boundaries.
  - **Destructive margins** (or convergent boundaries) usually involve an oceanic plate and a continental plate moving towards each other and colliding. The denser oceanic plate is forced beneath the less dense continental plate (a process known as **subduction**) and land is destroyed. The melting plate causes volcanic activity and the stress created by friction between the colliding plates can cause earthquakes.
  - **Conservative margins** are where plates may move in opposite directions to each other, or in the same direction, but at different speeds. As the plates move, their edges stick together, causing friction and stress to build up until one plate jolts forward. This sudden release of energy can cause earthquakes and rift valleys. There is no volcanic activity at conservative margins.

> ### Key Point
>
> The Earth is made up of:
>
> - inner **core**: solid iron and nickel, temperature around 5500°C
> - outer core: liquid iron and nickel
> - **mantle**: semi-molten rock (magma)
> - **crust**: thin layer of solid rock.

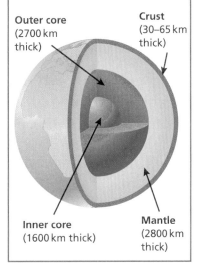

Outer core (2700 km thick)

Crust (30–65 km thick)

Inner core (1600 km thick)

Mantle (2800 km thick)

> ### Key Point
>
> Different physical processes occur at different types of plate boundary.

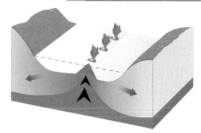

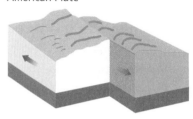

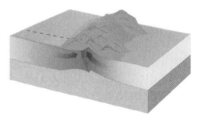

▲ Eyjafjallajökull eruption, Iceland, 2010
▲ Merapi eruption, Indonesia, 2010
☆ Japan earthquake, 2011
☆ Haiti earthquake, 2010

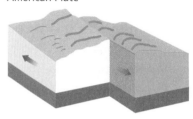

Constructive, e.g. Mid-Atlantic Ridge

Destructive, e.g. the Nazca Plate subducting beneath the South American Plate

Conservative, e.g. San Andreas fault

Collision zone, e.g. Himalayas and Alps

- Earthquakes (e.g. the 2015 Nepal earthquake) can also occur at **collision zones**. If continental plates of the same density collide, they buckle up rather than subduct. This creates fold mountains. There is no volcanic activity at collision zones.
- Most of the world's earthquake and volcano activity is found at the plate boundaries. '**Hot spots**' are exceptions and can occur at any location where the mantle is very close to the surface.
- Volcanic activity is common where the plates meet around the rim of the Pacific Ocean; the area is known as the 'Pacific Ring of Fire'.

Why People Still Live in High-Risk Areas

- They are near friends and family and have a job there.
- Confidence that their government has the ability to 'fix' things.
- Attitude of 'it won't happen to me – it's safe'.
- Volcanic areas have fertile soils, valuable minerals, geothermal energy and tourism opportunities.

Reducing the Risk and Effects

- The 3 Ps for locations close to earthquake and volcano zones are:
  - **Plan/predict**: It is possible to predict where earthquakes will occur, but not when. It is possible to predict volcanic eruptions; tell-tale signs include mini-earthquakes, escaping gas and a change in the volcano's shape. Governments can plan evacuations when volcanoes show signs of eruption.
  - **Protect**: Buildings and bridges can be designed to withstand earthquakes and strengthened to withstand the weight of volcanic ash. Firebreaks can reduce the spread of fire.
  - **Prepare**: Emergency services can be trained and prepared. People can be taught how to react in an earthquake or an evacuation. Countries can receive emergency aid or longer-term aid to help rebuild infrastructure and buildings.

Key Point

Monitoring, prediction and planning can reduce the risk from tectonic hazards.

Key Words

core
mantle
crust
plate margin
constructive margin
destructive margin
conservative margin
collision zone
hot spot

> ### Quick Test
>
> 1. Describe destructive and constructive plate boundaries.
> 2. Give two reasons why people still live near active volcanoes.

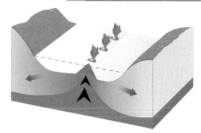

Revise

Tectonic Hazards 2

You must be able to:

* Understand how the effects of, and responses to, earthquakes vary between areas of contrasting levels of wealth.

Anatomy of an Earthquake

* Focus – point below surface where an earthquake occurs.
* **Epicentre** – the location on the surface directly above the focus.
* Seismic waves – shock waves of energy that travel through rock to the surface.
* Seismogram – measures shaking by an earthquake.
* Aftershocks – smaller tremors in the days after an earthquake.
* **Richter scale** – measures energy released in an earthquake.
* **Primary effects** – occur straight away when a tectonic hazard strikes.
* **Secondary effects** – occur later on, bringing more problems to those affected.

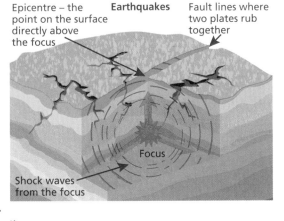

Epicentre – the point on the surface directly above the focus

Earthquakes

Fault lines where two plates rub together

Focus

Shock waves from the focus

Earthquake Case Study – Tohoku, Japan (Higher Income Country)

Background
* 11 March, 2011 – 2:46 pm.
* Epicentre in the ocean 62 miles (100 km) north-east of Honshu island.
* A complex destructive plate margin involving the Pacific, Okhotsk, Eurasian and Philippine plates.
* 9.0 on the Richter scale.
* The focus was 18 miles (29 km) below the surface.
* Fourth largest earthquake in the world.

Primary Effects
* Buildings destroyed and whole towns demolished.
* Massive damage in Sendai City (1 million people).
* Homes destroyed.
* Loss of power supplies.
* Roads blocked.
* Water shortages.
* Effects felt in Tokyo 185 miles (298 km) away.
* Japan's main island (Honshu) moved 7.8 feet (2.4 metres).

Secondary Effects
* Resulting **tsunami** (with waves up to 40 m high) killed 15 000 people.
* 2 million people left homeless.
* Layers of mud and debris left on land.
* Buildings, vehicles and bodies washed out to sea.
* Workers unable to shut down the Fukushima nuclear reactor, allowing radiation to leak.
* Japan's stock market collapsed.

> **Key Point**
>
> Higher income country is abbreviated to **HIC**.

Aftermath of the 2011 Japan earthquake

Immediate Responses

- Aid came from many other nations.
- A nuclear exclusion zone was established.
- Tsunami warning issued 3 minutes after the earthquake.
- Japanese Red Cross mobilised emergency teams.
- Warning given to other locations, such as Hawaii.

Long-Term Responses

- New, efficient tsunami warning system.
- Modifications to tsunami walls and floodgates that had not been effective.

Earthquake Case Study – Haiti (Lower Income Country)

Background

- 12 January, 2010 – 4:53 pm.
- Epicentre 10 miles (25 km) south-west of Port-au-Prince city.
- 7.0 on the Richter scale.
- Focus only 10–15 km below surface.
- Haiti is the poorest country in the Western Hemisphere.
- North American Plate slid past the Caribbean Plate.

Primary Effects

- 316 000 killed.
- 1 million homeless.
- Difficult to get aid into the country.
- Homelessness, loss of power and roads blocked.
- Airport, port and hospitals closed.

Secondary Effects

- 1.6 million people in refugee camps.
- Water and food shortages.
- Increase in crime, particularly looting.
- Outbreaks of cholera.

Immediate Responses

- Aid slow to arrive due to the damaged port and airport.
- The USA sent troops to support an aid programme.
- Many left Port-au-Prince city as their homes had been destroyed.

Long-Term Responses

- Dependence on overseas aid.
- $100 million aid from the USA and $330 million from Europe.
- New homes were built to a higher standard.
- Rebuilding of the port.

> **Key Point**
>
> Countries of contrasting levels of wealth show different effects and responses to tectonic hazards.

> **Key Point**
>
> Lower income country is abbreviated to **LIC**.

One of the tent cities in Haiti following the 2010 earthquake

> **Key Words**
>
> epicentre
> Richter scale
> primary effects
> secondary effects
> HIC
> tsunami
> LIC

> **Quick Test**
>
> 1. What is the difference between the primary and secondary effects of a tectonic hazard?
> 2. What is the difference between the focus and the epicentre of an earthquake?

Tectonic Hazards 3

You must be able to:

- Understand how the effects of, and responses to, volcanoes vary between areas of contrasting levels of wealth.

Features of a Volcano

- **Composite volcanoes** are made of alternating layers of lava and ash and are the result of multiple eruptions over hundreds of years. They give explosive eruptions of lava and ash. They are more likely to be found along destructive plate boundaries.
- **Shield volcanoes** form from runny magma (does not trap gases):
  - There is no build-up of pressure and eruptions are not explosive.
  - The runny lava flows a long way from the eruption, creating wide volcanoes.
  - They are more likely to be found on constructive plate boundaries, and at hot spots.
- **Pyroclastic flow** – torrent of hot ash, rocks, gases and steam, moving at up to 450 mph (700 km/h).
- **Ash cloud** – blocks out the Sun.
- **Lahar** – 60 mph mudslide of melted snow and volcanic ash.
- **Active volcano** – erupted recently.
- **Dormant volcano** – currently inactive but might erupt in future.
- **Extinct volcano** – not erupted for many thousands of years.

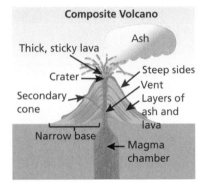

Composite Volcano

Thick, sticky lava — Ash — Crater — Steep sides — Vent — Secondary cone — Layers of ash and lava — Narrow base — Magma chamber

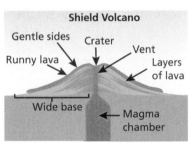

Shield Volcano

Gentle sides — Crater — Vent — Runny lava — Layers of lava — Wide base — Magma chamber

Volcano Case Study – Eyjafjallajökull, Iceland (Higher Income Country)

Background

- Erupted in April, 2010.
- Occurred at the spreading Mid-Atlantic Ridge of the constructive plate margin.
- North American Plate moving west; Eurasian Plate moving east.

Primary Effects

- Volcanic ash plume at 11 000 metres.
- Erupted under the glacier, causing severe flooding.
- Damage to roads, bridges, water supplies and livestock.
- Lava flows.
- Very fine-grained ash was a hazard to air traffic.

Secondary Effects

- Air space closed in Europe and thousands of flights cancelled.
- Glaciers covered in ash for months (increasing glacial melt).

Immediate Responses

- 500 farm families evacuated.
- National Emergency Agency dredged rivers, cleared ash and installed temporary bridges.

The ash cloud from the Eyjafjallajökull eruption caused the cancellation of flights all over the world

Long-Term Responses

- Individual nations grounded air traffic according to local weather conditions following the eruption.
- A scientific review investigated the effect of ash eruptions on air traffic.
- Further research continues to find better ways of monitoring ash concentrations and improving forecast methods.

Volcano Case Study – Mount Merapi, Indonesia (Lower Income Country)

Background

- Erupted in October, 2010.
- Caused by the Indo-Australian Plate subducting beneath the Eurasian Plate.
- Destructive plate margin in the 'Pacific Ring of Fire'.

Primary Effects

- 353 people were killed.
- 360 000 people were displaced to 700 emergency shelters.
- Volcanic bombs and hot gases for up to 11 km.
- Pyroclastic flows of up to 14 km.
- Ash fell up to 30 km away.
- Villages close to the volcano were buried.

Secondary Effects

- 350 000 people left homes in the area.
- Sulphur dioxide blown across Indonesia and as far as Australia.
- Ash cloud disrupted air transport.
- Roads blocked.
- Lahars.
- Food prices increased.
- Airports closed.

Immediate Responses

- Evacuation centres were established.
- 20 km exclusion zone was set up around the volcano.
- International aid arrived from charities like the Red Cross.

Long-Term Responses

- People moved to newer and safer homes.
- Funds were provided for farmers to replace livestock.
- Data from the eruption contributed to computer models to improve predictions.
- Improved warning and evacuation procedures.
- Construction of dams to hold back future lahars.

> **Quick Test**
>
> 1. Which has a wider base: a composite or a shield volcano?
> 2. What is the difference between pyroclastic flow and a lahar?

> **Key Point**
>
> The effects of and responses to volcanic eruptions vary between areas of contrasting wealth.

The eruption of Mount Merapi sent down tonnes of ash

> **Key Words**
>
> composite volcano
> shield volcano
> pyroclastic flow
> lahar
> active volcano
> dormant volcano
> extinct volcano

Tropical Storms

You must be able to:

- Describe and explain where tropical storms occur and why they are formed there
- Describe and explain the structure and features of a tropical storm
- Consider the possible links between climate change and tropical storms.

Global Distribution of Tropical Storms

- **Atmospheric circulation** is the movement of air around the Earth in cells (see page 28), transferring and redistributing energy. At sea level there are alternate low and high-pressure belts. It may be helpful to think of the atmosphere as an 'ocean' of air with waves, whirlpools and currents.

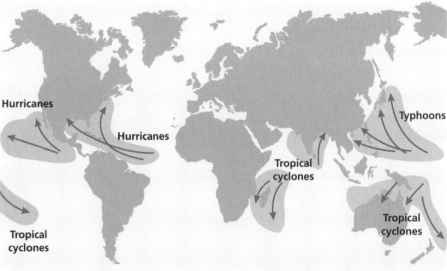

- **Tropical storms** are intense low-pressure systems, with distinct structure and features, formed over warm ocean waters in low latitudes. They are linked to the atmospheric circulation because warm, humid converging air at sea level starts to rise, creating low-pressure systems. If the average wind speed exceeds 74 mph (119 km/h), the storms are called 'cyclonic'.
- The Northern Hemisphere experiences more tropical storms and most occur between June and November. In the Southern Hemisphere, the tropical storms usually occur between November and April.
- Region-specific names for tropical storms are:
  - **Cyclone**: Indian Ocean and near Australia
  - **Hurricane**: Atlantic, Caribbean and Pacific (near North and South America)
  - **Typhoon**: Pacific (near Asia).

What Causes Tropical Storms?

- Tropical storms form over warm ocean water ($\geqslant$ 26–27°C) between latitudes 5° and 20° north and south of the Equator.
- Moist air evaporates and rises, creating low pressure.
- More warm, moist air is sucked in. A 'fountain' of air is formed.
- As it rises and cools, condensation occurs, creating clouds and convectional rain.

> ### Key Point
>
> Tropical storms are intense low-pressure systems, with a distinct structure and features, formed over warm ocean waters in low latitudes.

Hurricane Katrina, August 2005: a Category 5 Tropical Storm Which Devastated New Orleans in Southern USA

Anti-clockwise rotation in Northern Hemisphere

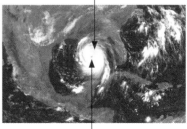

Eye

The Structure and Features of a Tropical Storm

- A tropical storm can be around 600 miles (1000 km) in diameter and has a distinctive structure.
- Owing to the Earth's rotation, **Coriolis force** occurs, which makes low-pressure systems spin anti-clockwise in the Northern Hemisphere and clockwise in the Southern Hemisphere.
- If a tropical storm becomes 'cyclonic', it spins so fast that the air around the centre forms a vortex, which creates an eye 20–40 miles (30–65 km) wide.
- The intense low pressure creates a 'dome' of seawater and a **storm surge** occurs, which can lead to coastal flooding.
- Cooler, drier air is dragged down into the eye. Consequently, inside the **eye of the storm** is a place with strangely calm weather and very little cloud or rain, whereas around the edge of the eye the winds are strongest.
- The Pacific Ocean has more severe tropical storms than any other basin.
- When a tropical storm makes landfall, it loses its energy supply (warm water) and slows owing to friction (especially if the land is hilly).

Climate Change and Tropical Storms

- Warm ocean water drives tropical storm formation.
- Warmer oceans expand, so storm surges may be worse.
- Climate change may alter the distribution of tropical storms and their **frequency** and **intensity** may increase, but the evidence for this is inconclusive.

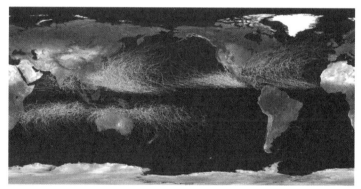

The cumulative tracks of all tropical storms for the period 1985–2005

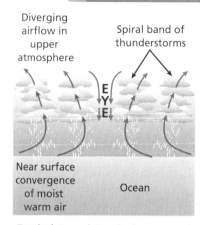

Diverging airflow in upper atmosphere

Spiral band of thunderstorms

Near surface convergence of moist warm air

Ocean

Tropical storm intensity is measured using the Saffir–Simpson scale

Five	≥70 m/s, ≥137 knots ≥157 mph, ≥252 km/h
Four	58–70 m/s, 113–136 knots 130–156 mph, 209–251 km/h
Three	50–58 m/s, 96–112 knots 111–129 mph, 178–208 km/h
Two	43–49 m/s, 83–95 knots 96–110 mph, 154–177 km/h
One	33–42 m/s, 64–82 knots 74–95 mph, 119–153 km/h

Key Point

If the oceans become warmer due to climate change, this may affect the distribution, frequency and intensity of tropical storms.

Key Words

atmospheric circulation
tropical storm
cyclone
hurricane
typhoon
Coriolis force (Coriolis effect)
storm surge
eye of the storm
frequency
intensity

Quick Test

1. Which hemisphere has the most tropical storms?
2. Which ocean has the highest number of severe tropical storms (≥ Category 4)?
3. Why might a storm surge cause more deaths than high winds?
4. Why do tropical storms weaken on landfall?
5. What is the difference between frequency and intensity?

Tropical Storms – Case Study

Quick Recall Quiz

You must be able to:

- Demonstrate your understanding of a case study of a tropical storm
- Describe and explain the primary and secondary effects
- Describe the immediate and long-term responses
- Evaluate some of the responses and explain how the 3 Ps can reduce the negative effects.

Case Study: Typhoon Haiyan

Context

- The Philippines in south-east Asia: an archipelago (island chain) made up of more than 7000 islands (around 4000 are inhabited)
- Population: 100 million (2015 estimate)
- Capital: Manila
- Area: 115 830 square miles or 300 000 km²
- Major religion: Catholicism (often a key player in supporting social infrastructure and resilience, especially in poorer communities)
- Average income: US $7500
- Typhoon Haiyan struck in November, 2013

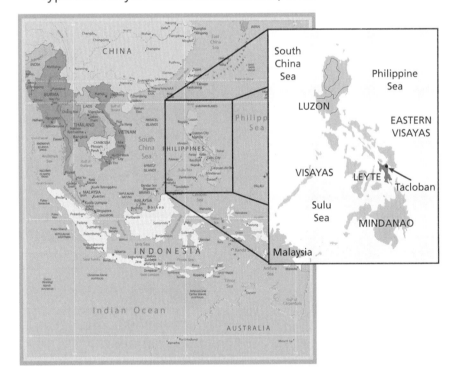

> **Key Point**
>
> Use the six Ws to structure your revision: Where? When? What? Who? Why? How?

The Storm and Why it Happened

- Typhoon Haiyan was a Category 5 tropical storm.
- It was the strongest storm ever recorded at landfall.
- Wind speeds were among the highest ever recorded in a tropical storm [1-minute sustained: 315 km/h (195 mph)].
- Worst storm in the Philippines for over 130 years.
- Warm surface water in the western Pacific.
- Climate change may have contributed.

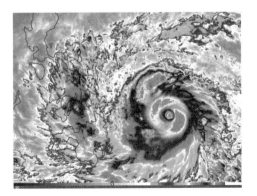
A satellite image of Typhoon Haiyan

Primary and Secondary Effects

- **Primary effects** cause widespread devastation – most settlements on the island of Leyte were destroyed.
- Landslides occurred across the landscape.
- Storm surges of 5–6 m on the islands of Leyte and Samar.
- Tacloban Airport terminal was destroyed.
- The first floor of the Tacloban City Convention Center, which was serving as an evacuation shelter, was submerged and many people were drowned.
- **Secondary effects** damage **infrastructure**, impeding relief efforts.
- Economic effects included high losses due to businesses being damaged or closed, and development was halted.
- Social effects included homelessness (1.9 million people); displacement (6 million people); bereavement; disease due to lack of food, water, shelter and medication; and schools closed.
- Ecosystems were damaged and farmland lost.
- Human factors made matters worse (**3 Ps**):
  - prediction: high-level warnings were too late
  - protection: communications infrastructure was too vulnerable (it failed in the Visayas islands)
  - planning: many people chose to stay in their homes (**inertia**).

Devastation after Typhoon Haiyan

Death Toll and Responses

- At least 6300 were killed (the exact number is unknown). The number of dead from the Eastern Visayas was 5877.
- Most of the dead were victims of the storm surge.
- Immediate responses:
  - Much of the central Philippines (Visayas) was placed under a state of national emergency.
  - Worldwide relief effort with aid of over $500 m (£400 m).
- Long-term responses: how did management strategies reduce the risk, and how can they do so in future?
  - The authorities were heavily criticised for being too reactive, rather than proactive.
  - Following a review of the 3 Ps, the authorities adopted much more proactive strategies with 'zero casualty' targets. They tackled the issue of inertia with incentives such as free bags of rice to persuade people to leave homes and property behind.
  - 'Evacuation rather than rescue, that's our doctrine,' said an emergency management chief.
  - The strategies were very successful: just over a year after Haiyan, Typhoon Hagupit hit the Philippines. It was also a Category 5 storm, but in marked contrast only 18 people were killed.

> ### Key Point
>
> The effects of Typhoon Haiyan (local name: Yolanda) on the Philippines were significantly affected by the country's **resilience**, which depends on its level of development and preparedness.

UK overseas aid workers loading emergency aid on to trucks after Typhoon Haiyan

> ### Key Point
>
> Be aware that hazardous events occur on a frequent basis and you should have knowledge of up-to-date examples.

> ### Key Words
>
> infrastructure
> resilience
> 3 Ps
> inertia

Quick Test

1. Which country was most severely hit by Typhoon Haiyan?
2. Give a measure of its magnitude.
3. Describe one primary and one secondary effect.
4. Give one of the causes.
5. Outline one immediate and one long-term response.

Quick Recall Quiz

Extreme Weather in the UK

You must be able to:

- Demonstrate that you can explain the main influences on the UK climate
- Describe the main types of atmospheric hazards affecting the UK
- Analyse the causes, impacts and responses of an extreme weather event in the UK.

UK Climate and Weather Hazards

- The UK has a temperate climate, i.e. relatively moderate temperatures and levels of precipitation.
- A key factor is its location in the **mid-latitudes** on the coast of north-west Europe facing the Atlantic, where cold and warm **air masses** create **fronts**, which produce precipitation.
- The **prevailing winds** (most common winds) are westerlies from the Atlantic Ocean, and the relatively warm, moist air from the **North Atlantic Drift** (or **Gulf Stream**) creates a dominant **maritime** effect.
- Less frequently, the UK receives relatively dry **continental** air from land masses to the south and east. In winter, such air can be extremely cold but in summer it is very hot.
- **Altitude** is also influential. Higher ground in the west has greater precipitation (**relief rainfall**) and lower temperatures. Consequently, the east is in a **rain shadow** (receiving much lower precipitation).
- The main hazards affecting the UK are atmospheric or **hydro-meteorological hazards** linked to the water cycle.
- The main types of weather hazard event include: river flooding; sea flooding; winter storms; snow and ice; **drought**.

> ### Key Point
>
> The main weather hazards impacting the UK include river and sea flooding, winter storms, snow, ice and drought.

The UK Drought of 2010–12

- Droughts are extended dry periods lasting for months or years.
- The 2010–12 drought was one of the ten most significant for 100 years.
- The pattern of such events may contribute to evidence that weather is becoming more extreme in the UK due to climate change.

Causes

- **Blocking highs** (slow-moving anticyclones) led to very dry winters in 2009–10 and 2011–12.
- East winds were common, bringing in dry continental air.
- A significantly lower precipitation than normal (see map, right).

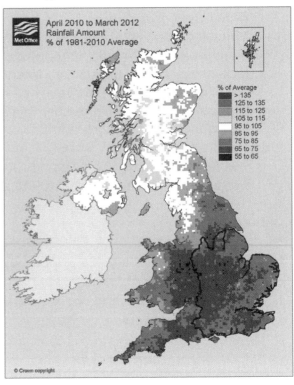

April 2010 to March 2012
Rainfall Amount
% of 1981–2010 Average

% of Average
> 135
125 to 135
115 to 125
105 to 115
95 to 105
85 to 95
75 to 85
65 to 75
55 to 65

© Crown copyright

- Climate change may have contributed.
- An imbalance between the demand and supply of water in the densely populated south-east and Midlands, which suffer **water deficit** and **water stress**. These areas rely on piping of water from regions of **water surplus** and/or groundwater supplies.

Impacts

- Economic and social impacts:
  - farmers struggled to provide water for livestock and to harvest crops
  - low reservoir levels.
- Environmental impacts:
  - groundwater and river levels were very low, affecting aquatic ecosystems
  - wildfires spread.

Management Strategies

- Hosepipe bans were introduced (affecting six million consumers).
- Water meters were installed to monitor usage.
- Water companies fixed leaking pipes.
- Water-saving devices were encouraged.
- Education on water use.
- Improved waste-water recycling.
- New reservoirs and pipelines were considered.
- Desalination plants were being considered (but at high cost).

Key Point

Extreme weather events such as drought have economic, social and environmental impacts.

Key Words

mid-latitudes
air mass
front
prevailing wind
North Atlantic Drift
Gulf Stream
maritime
continental
altitude
relief rainfall
rain shadow
hydro-meteorological
 hazard
drought
blocking high
water deficit
water stress
water surplus

Quick Test

1. What is meant by a maritime influence on the UK climate?
2. a) Give an example of one extreme weather event in the UK.
   b) Describe one impact and one management strategy for the extreme weather event.

Climate Change

You must be able to:

- Describe and explain the main evidence for, and causes of, climate change
- Assess different approaches to the management of climate change.

Evidence for Climate Change

- The Intergovernmental Panel on Climate Change (**IPCC**) is an internationally accepted authority on climate change.
- The IPCC found that Northern Hemisphere temperatures during the 20th century were the highest for the past 1300 years.

Long-Term Evidence

- In the absence of reliable climate data, **proxy measures** are used, e.g. ice cores, marine sediment cores and pollen analysis.
- Proxy measures of temperature since the start of the **Quaternary period** (the last 2.6 million years) show that the Earth has experienced an 'Ice Age' of several lengthy **glacials** (extremely cold periods of glacier growth) interspersed with shorter, warmer **interglacials** (warmer periods of glacier retreat).
- The Earth is currently in an interglacial period.

Short-Term Evidence (Last Few Hundred Years)

- As climate measurements were unreliable before 1850, other proxy measures are needed such as tree rings and historical sources (e.g. landscape paintings and literature).
- Data show a marked global temperature rise since 1850 – the 'hockey stick' graph shows variations in the Northern Hemisphere temperature over the last 1000 years.
- Warming oceans, sea-level rises, reduction in Arctic sea ice, increases in the frequency and intensity of extreme weather events, are all evidence of climate change.
- The atmosphere creates a natural **greenhouse effect**, allowing for short-wave solar radiation (e.g. in light), but **greenhouse gases** prevent the escape into space of some of the heat created (long-wave infrared radiation).
- Industrialisation and population growth have increased greenhouse gas emissions.
- Carbon dioxide increase was first noticed in the 1950s and 60s by scientists in Hawaii, with record highs in 2021.
- Each year since 2015, global temperatures have been more than 1°C above the 'pre-industrial' period of 1850–1900.
- 2013–2021 all rank among the ten warmest years on record.

The 'Hockey Stick' Graph Showing Variations in the Northern Hemisphere Temperature Over the Last 1000 Years

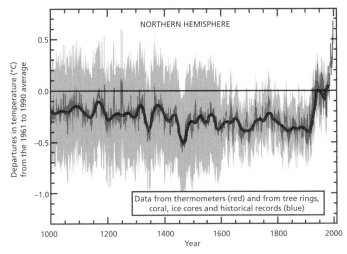

NORTHERN HEMISPHERE

Data from thermometers (red) and from tree rings, coral, ice cores and historical records (blue)

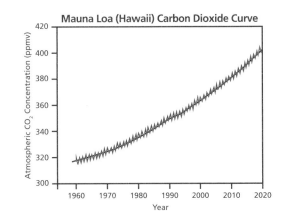

Mauna Loa (Hawaii) Carbon Dioxide Curve

Climate Change Causes

- Natural factors include:
  - orbital changes (Milankovitch cycles)
  - volcanic activity (volcanic emissions block sunlight)
  - solar output (changes in the Sun's energy).
- Human factors (**anthropogenic** factors) include:
  - the use of fossil fuels producing greenhouse gases, such as power generation and transportation
  - agriculture, such as methane from cattle and rice paddies
  - deforestation, e.g. tree loss reduces natural **carbon capture** (**carbon sequestration**)
  - methane release from melting permafrost and ocean floors, e.g. due to anthropogenic global warming.

Managing Climate Change Impacts

- The IPCC recommends **mitigation** and **adaptation**.
- Mitigation means cutting greenhouse gas emissions, reducing or eliminating long-term risk to human life and property, internationally agreed targets (such as the Paris Agreement ratified in November, 2016) to reduce greenhouse gas emissions, alternative methods of energy production and carbon capture.
- Carbon capture removes carbon at the emission source and stores it deep underground.
- Adaptation means making adjustments to reduce potential damage, to limit the impacts and to take advantage of new opportunities, e.g. reducing the risk from rising sea levels and making changes to agricultural systems.
- Critics of mitigation say that reversal of climate change is not possible. A minority still deny climate change completely.
- Critics of adaptation say that without mitigation it only treats the symptoms, not the causes, of climate change.

Cuts to greenhouse gas emissions are a key part of tackling climate change

Quick Test

1. Give two examples of proxy measures.
2. Describe and explain one physical and one human cause of climate change.
3. Which responses are **not** adaptation? Which responses are **not** mitigation?

Climate change response	NOT adaptation	NOT mitigation
Electric cars		
Higher sea walls		
Tidal power		
Wind farm		
IPCC carbon reduction targets		
Improving air-conditioning in houses		

> **Key Words**
>
> IPCC
> proxy measure
> Quaternary period
> glacial
> interglacial
> greenhouse effect
> greenhouse gas
> anthropogenic
> carbon capture / carbon sequestration
> mitigation
> adaptation

Tectonic Hazards 1

1   What causes the Earth's plates to move?   [2]

2   Draw a diagram to help explain what happens at a destructive plate margin.   [4]

3   Draw a diagram to help explain what happens at a constructive plate margin.   [4]

4   Draw a diagram to help explain what happens at a conservative plate margin.   [4]

Total Marks _____ / 14

Tectonic Hazards 2

1   What do the terms LIC and HIC mean?   [2]

2   With reference to the Tohoku, Japan 2011 earthquake, which statements are true and which are false?

     A   A nuclear power station was seriously damaged.
     B   The epicentre was at sea.
     C   It measured over 6 on the Richter scale.
     D   It produced tsunami waves up to 40 metres in height.
     E   The tsunami waves only affected a nuclear reactor.   [5]

3   Explain why the largest earthquakes do not always result in the most deaths.   [6]

Total Marks _____ / 13

Tectonic Hazards 3

1   Describe the effects of the Mount Merapi volcanic eruption in 2010.   [4]

2   Identify two primary effects following the eruption of Eyjafjallajökull.   [2]

3   What long-term responses can be adopted to reduce the effects of a volcanic eruption?   [4]

Total Marks _____ / 10

Tropical Storms

1 Which of the following statements is true?

A Tropical storms form along the Equator.
B Tropical storms form between latitudes 5° and 20° north and south of the Equator.
C Tropical storms form between latitudes 30° and 50° north and south of the Equator.
D Tropical storms form between latitudes 40° and 60° north and south of the Equator.
E Tropical storms form between latitudes 50° and 70° north and south of the Equator. [1]

2 Give three key features of a tropical storm, such as the one below. [3]

3 What are the region-specific names used for tropical storms in these areas?

a) Indian Ocean and near Australia
b) Atlantic, Caribbean and Pacific (near North and South America)
c) Pacific (near Asia) [3]

4 How does the Coriolis force (Coriolis effect) affect wind direction around tropical storms in different parts of the world? [4]

Total Marks .... / 11

Tropical Storms – Case Study

1 What are the '3 Ps'? [3]

2 For a tropical storm you have studied, state its name, where and when it occurred. [3]

3 Select the main category for each of these impacts of a tropical storm:

Impacts	Economic	Social/Political	Environmental
Homelessness			
Factories and other businesses closed or inaccessible due to damage to transport infrastructure			
Waterborne diseases			
Damage to ecosystems			
Schools closed for weeks			

[5]

4 For these effects of a tropical storm, indicate which are primary (P) and which are secondary (S).

a) People drowned in storm surge.

b) Thousands of people are homeless for over a year.

c) Outbreaks of cholera and dysentery kill hundreds of people.

d) Severe damage to infrastructure, such as railways and bridges, by flash floods.

e) Homes destroyed by high winds.

f) Schools and businesses are closed for months. [6]

Total Marks _____ / 17

Extreme Weather in the UK

1 For an extreme weather event in the UK that you have studied, state when it occurred. [1]

2 How does altitude of the land influence the UK climate? [2]

3 What are the main types of weather hazard affecting the UK? [3]

4 Outline the main causes of an extreme weather event in the UK. [4]

Total Marks _____ / 10

Climate Change

1 IPCC is the abbreviated name of which organisation?

    A International Policy on Climate Change
    B International Panel on Climate Change
    C Intergovernmental Policy on Climate Change
    D Intergovernmental Panel on Climate Change
    E Intermediate Panel on Climate Change   [1]

2 What key term is sometimes used to describe human causes of climate change?   [1]

3 Give three human causes of climate change.   [3]

4 Draw lines to match the key terms to the correct definitions.

Key term	Definition
Glacial	Indirect ways to find out average temperatures from the past
Interglacial	Last 2.6 million years, including the 'Ice Age'
Quaternary	Warmer periods of glacier retreat
Proxy measure	Extremely cold periods of glacier growth

[3]

Total Marks .... / 8

Ecosystems and Balance

You must be able to:

- Describe how ecosystems are structured
- Explain, using a small-scale UK example, how the balance of ecosystems can be affected by a change to one component.

What is an Ecosystem?

- **Ecosystems** are communities of living and non-living components that all function together to create a distinctive environment.
- **Food chains** show simple relationships between different organisms. Basically that means, what eats what.

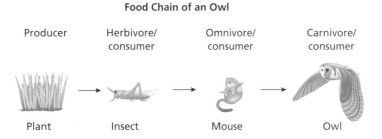

Food Chain of an Owl

Producer	Herbivore/ consumer	Omnivore/ consumer	Carnivore/ consumer
Plant	Insect	Mouse	Owl

- **Food webs** show more complex **interrelationships** between organisms.

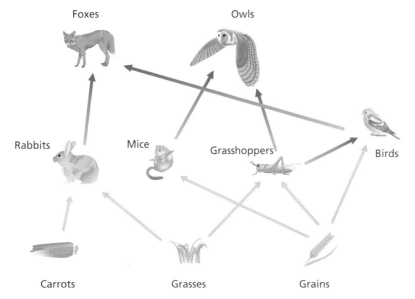

- Ecosystems can be small scale, such as a pond or hedgerow in the UK.
- Ecosystems can be large scale, such as a tropical rainforest, and these are known as 'biomes'.
- Ecosystems are balanced – each organism plays its part and can be easily disrupted, particularly by the actions of humans.

> ### Key Point
>
> Ecosystems are a critical part of the natural world. All species alive today form part of an ecosystem. The climate and soil form the building blocks of all ecosystems.

A pond is a small-scale ecosystem

- **Producers** are organisms such as plants that convert the Sun's energy into sugars, thus *producing* food for them.
- **Consumers** are species that eat other species. They can be **herbivores** (such as goats) that eat plants or **carnivores** (such as lions) that eat other animals.
- **Decomposers** such as fungi act to break down the remains of dead plants and animals and then return their nutrients to the ecosystem.

Balance in Ecosystems

- Maintaining balanced ecosystems is critical to the survival of life on Earth.
- A hedgerow is one small-scale example of an ecosystem in the UK.
- In a hedgerow, the plants (such as small trees) are the producers, small animals (such as mice) are herbivore consumers and birds (such as owls) are carnivore consumers.
- Ecosystems naturally manage themselves in a variety of complex ways: populations of particular organisms are kept in check by being the prey of another – too many beetles born in one year means more food for the birds that prey on them.
- Nutrients move around an ecosystem through the **nutrient cycle** – plants grow and are eaten by herbivores; the herbivores are then eaten by carnivores; the carnivores die and decompose; the decomposed remains are reabsorbed by the trees. At each stage, nutrients are transferred around the ecosystem.
- Every part of the Earth's surface is part of an ecosystem. The important ecosystems for this exam are tropical rainforests, hot deserts, polar and tundra.
- Human activity can disrupt the balance of an ecosystem. An example of this is deforestation, as the removal of trees destroys the habitats of countless organisms, thus leading to potential imbalance. The logical endpoint is the disappearance of a particular ecosystem.

A hedgerow is another type of small-scale ecosystem

Key Point

An ecosystem functions properly when all the parts of it work in harmony. This harmony can be altered by human actions.

Key Words

ecosystem
food chain
food web
interrelationship
producer
consumer
herbivore
carnivore
decomposer
nutrient cycle

Quick Test

1. Give an example of a large-scale and a small-scale ecosystem.
2. What is the difference between a food web and food chain?
3. Suggest what might happen if one species reproduces too much.
4. How do nutrients move around an ecosystem?

Ecosystems and Global Atmospheric Circulation

You must be able to:

- Describe how global atmospheric pressure changes between January and July
- Describe the locations of a range of global ecosystems.

Global Atmospheric Circulation

- Differences in **atmospheric pressure** exist. Atmospheric pressure fluctuates constantly but patterns of air movement can be observed.
- Pressure patterns can be broadly predicted and follow seasonal trends.
- Wind is the movement of air owing to many factors, such as the movement of air masses.
- Air masses can be disrupted and their behaviour distorted by features such as mountains, coasts and rivers.
- **Wind patterns** can be predictable but may be extremely erratic.
- The locations of the world's major ecosystems are heavily influenced by the existence and movement of the Earth's major high and low-pressure areas and the resulting wind patterns.

> **Key Point**
>
> Atmospheric circulation affects where ecosystems are found.

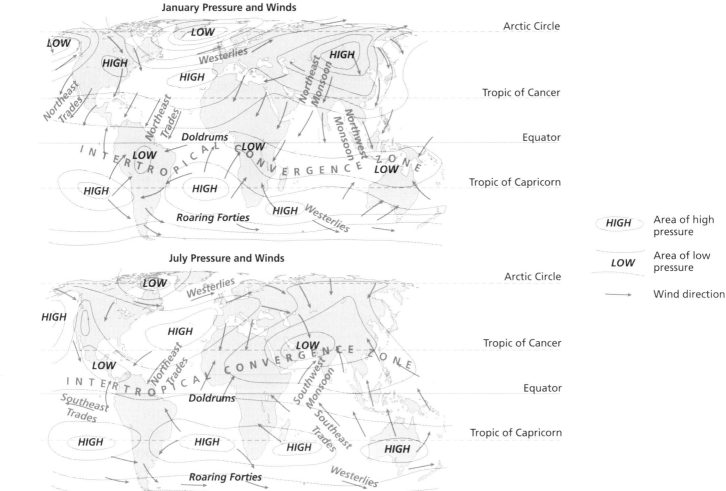

Global Ecosystems

Tropical rainforests
- Found in a wide belt encircling the Earth, roughly following the Equator and for the most part between the Tropics of Cancer and Capricorn.
- The Amazon, in the northern part of the Southern Hemisphere, is the world's largest rainforest.
- The climate is hot and wet all year round.

Tropical grasslands
- Named after the Tropics of Cancer and Capricorn, along both of which they can be found.
- The most famous tropical grasslands are those of the savannas of central Africa.
- They have long dry and brief wet seasons.
- They have little tree cover.

Temperate grasslands
- Found in two wide belts to the north of the Tropic of Cancer and to the south of the Tropic of Capricorn.
- They flourish in the centres of the continents of North America (prairies) and Asia (steppe).
- They have warm summers with very cold winters.
- They have little tree cover.

Temperate forests
- Found mostly in the Northern Hemisphere, well to the north of the Tropic of Cancer but to the south of the Arctic Circle.
- Examples of note are the deciduous forests of north-western Europe.
- They have warm, damp summers and mild winters.
- They have deciduous trees.

Boreal forests (taiga)
- Found in a thin belt just to the south of the Arctic Circle, fringing tundra-type ecosystems.
- Examples of places with boreal forests include central Canada.
- They have mild summers and very cold winters.
- The trees are mostly coniferous.

Tundra
- Ecosystems found in a wide belt encircling the Earth in what are known as the high latitudes – areas on and a little south of the Arctic Circle.
- Much of northern Russia, Canada and Iceland have tundra-type environments.
- They are very cold regions.
- They have little tree cover.

Polar regions
- Found in the far north, above the Arctic Circle, and far south, below the Antarctic Circle.
- Greenland in the north and Antarctica in the south are examples of polar regions.
- The climate is very cold all year round, but with little precipitation.

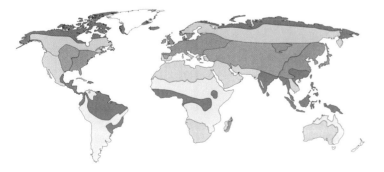

☐ Tropical rainforest
☐ Tropical grassland
☐ Hot desert (see pages 30-31)
☐ Temperate grassland

☐ Temperate forest
☐ Boreal forest
☐ Tundra
☐ Polar regions

Quick Test

1. Where are tropical rainforests found?
2. What things can affect the behaviour of air masses?
3. Give examples of places that have temperate forests.

 Key Words

atmospheric pressure
wind patterns

Rainforests and Hot Deserts – Characteristics and Adaptations

You must be able to:

- Describe the distinctive characteristics of rainforests and hot deserts
- Describe how plants and animals adapt to named rainforest and desert ecosystems.

Characteristics of Tropical Rainforests

- The climate in tropical rainforests is generally warm all year round, with average daily temperatures of around 27–29°C. Rainfall is also high at around 500–600 mm each month.
- The soil layer is known as a **latosol**. Latosols are deep but have few nutrients within them. The vast majority of the nutrients are found in the **leaf litter**, which decays rapidly due to the warm, damp climate.

Adaptations in the Rainforest

- Biodiversity (the range of species) is very high in rainforests but is at risk from human impact.
- Trees such as the mahogany and kapok in the Amazon have developed huge **buttress roots** that provide stability but are also largely above ground so as to absorb nutrients directly from the fast-decaying leaf litter.
- Pitcher plants have adapted to the low nutrient soils by developing a taste for insects – they are attracted using scent glands and then caught using a slippery flower before being digested.
- Forest elephants eat clay from ponds in rainforest clearings to counteract the toxins in the leaves they eat.
- Ecosystems are balanced and display interdependence – animals, plants, climate and water all play their part and can be disrupted by the actions of humans.

Characteristics of Hot Deserts

- The temperature in deserts is high all year round with summer temperatures reaching as high as 40°C and winter temperatures reaching up to 20°C.
- Monthly rainfall is low with an average of approximately 3 mm.
- Species have developed ways of dealing with the conditions.
- Rivers do flow through hot deserts – such as the Colorado in south-western USA – and humans in richer countries have sought to use them to help farm the land through **irrigation**.
- Rivers appear and disappear quickly and often as the result of high rainfall in mountain sources.

Tropical Rainforest Structure

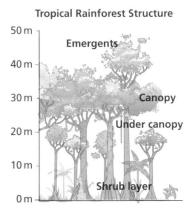

Buttress roots

 Key Point

Tropical rainforests are found along the Equator and between the Tropics of Cancer and Capricorn. The soils, plants and animals in a tropical rainforest have all developed to work together in harmony.

- Water in some deserts appears as snow, which some animals, like the Bactrian camel, have adapted to by eating the snow.
- The soil is poor and supports little vegetation.
- Over-irrigation can cause **salinisation**.
- The ecosystem is balanced – animals, plants, climate and *especially* water are all critical and easily disrupted by humans.

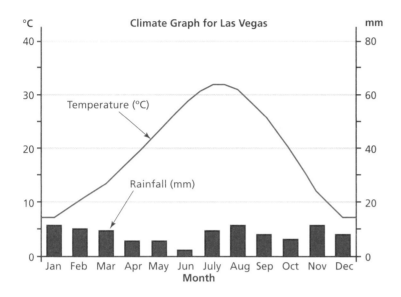

Climate Graph for Las Vegas

Fennec foxes

Adaptations in Hot Deserts

- The saguaro cactus of the Mojave Desert in North America has no leaves in order to cut down moisture loss through **transpiration**. It also has shallow but wide-ranging roots so that it can take advantage of brief rainstorms.
- The quiver tree of southern Africa has fleshy leaves that allow it to retain moisture.
- Fennec foxes are nocturnal, which allows them to avoid the **extreme temperatures** of the day. The fennec also has large ears, which act like a radiator in a car – letting heat escape.
- Biodiversity is low due to extreme temperatures and limited water.

> **Key Point**
>
> Deserts are found in different parts of the world for different reasons but they all share the same characteristics of extremely low rainfall and high summer daytime temperatures. Deserts have a low biodiversity.

> **Key Words**
>
> latosol
> leaf litter
> buttress roots
> irrigation
> salinisation
> transpiration
> extreme temperature

Quick Test

1. Describe the climate in a tropical rainforest.
2. Name two rainforest plants and their adaptations.
3. Describe the climate in hot deserts.
4. Name two hot desert plants and their adaptations.
5. Name two examples of hot deserts.

Opportunities, Threats and Management Strategies in the Amazon

You must be able to:

- Using case study evidence, describe opportunities, threats and management strategies in tropical rainforests such as the Amazon in South America.

Opportunities and Threats

- Tropical rainforests provide goods and services to humans:
  - Small groups of people called **hunter-gatherers** collect edible plants and catch wild animals to eat
  - Soils can be fertilised through small-scale **shifting cultivation** of crops such as manioc and cassava
  - Trees provide fuel and building materials
  - Medicines and hunting poisons can be extracted from a wide variety of plants and animals.
- Trees absorb atmospheric carbon and release oxygen.
- **Commercial** and **subsistence farming** of crops (such as soybeans) and animals (such as cows) is one of the major reasons for the deforestation of the Amazon.
- There are large deposits of minerals such as iron, bauxite, nickel and copper beneath the forest but to reach them means large-scale removal of the vegetation.
- Building roads such as the Trans-Amazonian Highway has led to increased **settlement** as Brazil's population has grown. Roads also allow access to harder-to-reach areas, thus creating new centres of expansion.
- **Hydroelectric power (HEP)** schemes such as that at Tucuruí in Brazil create huge amounts of energy for a growing economy but also flood huge areas of forest. The dams also encourage further deforestation for farming and settlement.
- Removing the vegetation causes soil erosion as roots that bound the soil together no longer do so, leaving it exposed and weakened.
- Deforestation also removes a valuable source of humus for the soil, leaving it unproductive for farming.
- With deforestation comes a loss of habitats and a resultant drop in biodiversity as species simply die out. Studies have also shown that large-scale deforestation can reduce rainfall rates.
- The fastest way to clear forest is through burning it, but this has led to huge amounts of carbon dioxide being released into the atmosphere, thus contributing to global climate change.
- Brazil has experienced positive **economic development** as dams have provided energy for industry and homes. Large-scale mining and farming have provided jobs to millions and given Brazil a source of export income.

Deforestation

Sustainable Management

- Rainforests such as the Amazon can be sustainably managed through some of the following measures:
  - Selective cutting (logging), whereby only trees above a certain height are felled, thus leaving smaller trees to attain maturity.
  - International agreements that name and make illegal the export of endangered plant and animal species. For example, the 1973 'Convention on International Trade in Endangered Species of Wild Fauna and Flora', better known as CITES.
  - Encouraging **ecotourism**, which makes the forest itself the tourist attraction. Tourists are shown and educated on the wonder and diversity of the ecosystem. Local populations can gain employment from such schemes, which in turn enriches the local economy without destructive exploitation and discourages **out-migration**.
  - Using labelling schemes such as the international Forest Stewardship Council (FSC), where trees grown in sustainable forests are marked as such, thus encouraging their purchase over trees felled either unsustainably or illegally.
  - International debt reduction schemes for poorer countries in order that endangered trees and animals are not seen as an easy source of export earnings, to help pay off debt.

Ecotourism can make an attraction of rainforests.

Clear Cutting

Forests

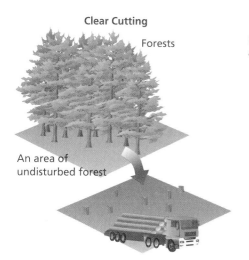

An area of undisturbed forest

In clear cutting, a large area of the forest is cleared, harming the forest. The forest is systematically cleared of all vegetation.

Selective Cutting

Non-valuable trees are left undisturbed

Commercially valuable trees are logged

In selective cutting, some trees are retained to minimise the damage caused by logging. Some trees are left undisturbed while commercially valuable ones are logged.

Quick Test

1. Give three reasons why deforestation has taken place in the Amazon.
2. Describe the impacts of deforestation.
3. How can humans use tropical rainforests without destroying them?

Key Words

hunter-gatherer
shifting cultivation
commercial farming
subsistence farming
settlement
hydroelectric power (HEP)
economic development
ecotourism
out-migration

Opportunities, Threats and Management Strategies in Hot Deserts

You must be able to:

- Using case study evidence, describe opportunities, threats and management strategies in hot deserts such as the Mojave in south-western USA or Almería in southern Spain.

Opportunities and Threats in the Mojave, South-Western USA

- In richer countries, governments build roads and bridges (**infrastructure**) that encourage new communities.
- Extensive **commercial farming** of crops such as peanuts has been made profitable by the building of the Hoover Dam (below) in Nevada, USA.
- The dam both controlled the Colorado River and created Lake Mead, which today is the source of water used for **irrigation**.
- The Hoover Dam also exists as a source of hydroelectric power (HEP).

Las Vegas

- Extreme temperatures and inaccessibility pose huge challenges to development in the Mojave.
- Tourism has become the biggest industry in the Mojave, for example the casinos in Las Vegas.
- Increased **desertification** is a real problem in places like the Mojave and other semi-arid places such as southern Spain.
- Growing demand for water from agriculture, tourism and settlement means less for the environment.
- Some communities are having their water shut off for parts of the day and farmers do not always have enough water to irrigate their crops or water their stock. Levels are so low in some reservoirs that it is threatening HEP generation.

> ### Key Point
>
> In the deserts of south-western USA, large-scale settlement has been made possible by the development of huge water projects such as the Hoover Dam on the Colorado River. By using dams to overcome the challenges of the environment, many new challenges have been created.

- Desertification has been accelerated across the globe by climate change, population growth, overgrazing by farmed cattle and over-cultivation.
- Strategies to deal with desertification in the Mojave include:
  - encouraging less water use among local people
  - developing planning laws that restrict the size of buildings
  - planting trees to stabilise sand dunes
  - encouraging the use of drip irrigation in farming.

Desertification in Almería, Spain

- Global warming has led to an increase in temperature and lower rainfall in Almería, southern Spain.
- **Pastoral** sheep farming has led to the removal of vegetation cover through overgrazing and trampling of the soil.
- **Terraced fields** created useful flat land for cultivation and rainfall by being built near natural watercourses.
- Many farms have been abandoned because of migration to the cities.
- Left untended, the terraces collapse, leaving the soil exposed to the Sun.
- Mining of gypsum – a mineral important to the building industry – has led to a reduction of underground **aquifers**.
- **Mass tourism** has increased the need for water. Water is needed for swimming pools, showers and for maintaining golf courses.
- Over-cultivation can lead to the process of salinisation.
- Salinisation occurs when the water in soils evaporates in high temperatures, drawing salts to the surface which are toxic to many plants.
- Strategies to deal with desertification in southern Spain include:
  - encouraging the use of drip irrigation
  - applying **appropriate technologies** that are cheap and easy for local farmers, such as applying a coating of almond shells to the soil
  - recycling water within tourist areas
  - controlling the size and number of golf courses.

View over Cortijo Grande Golf Course, Almería, Spain

Desert Tabernas, Almería, Spain

Quick Test

1. Give three reasons why desertification has increased.
2. How does salinisation occur?
3. How can humans deal with the threat of desertification?

Key Words

infrastructure
commercial farming
desertification
pastoral farming
terraced field
aquifer
mass tourism
appropriate technology

Polar and Tundra Environments

Quick Recall Quiz

You must be able to:

- Describe the physical characteristics of, and the adaptation of, species to polar and tundra environments and development opportunities therein
- Describe the challenges of developing a cold environment such as Antarctica and strategies to protect it, while balancing the needs of development.

Physical Characteristics of Polar and Tundra Environments

- Polar and **tundra** environments are both found in what are known as the high latitudes. Tundra is mostly found about 60–70° north.

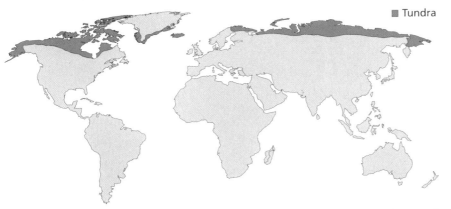

■ Tundra

- Climates in polar and tundra environments vary throughout the year, with places like Alaska and Norway receiving moderate summers that can present daytime temperatures of 20–25°C. Winter temperatures can plunge below –40°C.
- The characteristic landscape in tundra regions is that of **permafrost**: land that has been frozen for two or more consecutive years.

> ### Key Point
>
> Polar and tundra are both regarded as extreme environments. Both are cold and have characteristic ecosystems and climates. There are development opportunities present in both of these cold environments.

Adaptations

- Soil in tundra regions may remain frozen all year round. Frost-resistant plants such as the Arctic poppy survive by developing adaptations such as shallow roots and flowers that track the path of the Sun.
- Biodiversity in tundra regions is very low due to extreme cold.
- Permafrost, soil, plants and animals are all interdependent and at risk of change due to human actions.
- Animals such as the Arctic fox develop thick coats to protect against the cold.
- The Arctic hare has small ears to reduce heat loss and white fur to avoid the gaze of predators.
- There are huge deposits of minerals such as bauxite, tar sand and standard crude oil. Less profitable sources of oil, such as tar sand, have begun to be developed.

Arctic poppies

- The extraction of tar sand oil creates huge numbers of jobs but extraction destroys habitats.
- Polar ocean regions have rich seas, brimming with fish, and this attracts large-scale industrial fishing. An example is the Bering Sea near Alaska.

Antarctica – Development and Conservation

- Antarctica is the Earth's most southerly continent. There are no permanent settlements, although scientists from many countries occupy temporary stations in various places.
- Antarctica is classed as a desert due to its spectacularly low level of precipitation.
- Antarctica has huge mineral deposits such as coal, bauxite and crude oil.
- Antarctica is highly inaccessible as there is no infrastructure, no means of growing crops to sustain a population and any buildings would have to be built to withstand the extreme climate.
- Commercial fishing is big business in Antarctica, with crews from around the world (but especially Argentina and Chile) making large profits on huge catches. Governments must consider fishing quotas in order to balance the needs of fishermen and the protection of biodiversity.
- **Extreme tourism** is a recent development because tourists are being increasingly attracted to new and often hostile places.
- International pressure groups, such as Greenpeace, campaign for the protection of Antarctica.
- The 1959 Antarctica Treaty is an agreement by 12 countries declaring Antarctica a scientific preserve.
- The Madrid Protocol prohibits all mining in Antarctica to preserve it as a wilderness area.

Tar sand

> ### Key Point
>
> Antarctica is the world's last true **wilderness** and, as such, is in need of protection. Development opportunities exist but must be carefully weighed against the need to conserve the fragile habitats and ecosystems found within Antarctica.

> ### Quick Test
>
> 1. Describe the climate of polar and tundra regions.
> 2. What is permafrost?
> 3. What are the benefits and drawbacks of extracting oil from tar sands?
> 4. What is extreme tourism?

> ### Key Words
>
> tundra
> permafrost
> wilderness
> extreme tourism

Review Questions

Tectonic Hazards 1

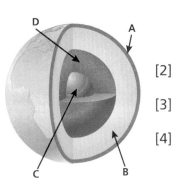

1 What is a 'hot spot'? [2]

2 State three reasons why tectonic hazards have a greater impact in LICs. [3]

3 Identify the different layers of the Earth labelled in this diagram. [4]

4 Describe how the '3 Ps' (predict, protect and prepare) can help to reduce the effects of an earthquake. [6]

Total Marks _____ / 15

Tectonic Hazards 2

1 What name is given to the standard scale used to measure energy release in an earthquake? [1]

2 In what ways may the effects of an earthquake in a higher income country be different to those experienced in a lower income country? [6]

Total Marks _____ / 7

Tectonic Hazards 3

1 Draw a diagram to help describe and explain the shape of a shield volcano. [4]

2 Match up the following key words with the correct definitions.

Key term	Definition
Ash cloud	Torrent of hot ash, rock, and gases and steam
Lahar	Blocks out the Sun, causing suffocation and health problems
Pyroclastic flow	Volcano nobody expects to erupt ever again
Extinct volcano	Volcano that has erupted in past 2000 years but is not currently active
Dormant volcano	Mudslide including rock debris and water

[4]

Total Marks _____ / 8

Tropical Storms

1 How fast does the wind speed in a tropical storm need to be for it to be called 'cyclonic'? [1]

2 Explain why coastal flooding may be caused by tropical storms. [3]

3 Explain how climate change may influence the occurrence of tropical storms. [3]

4 Complete the table, using these terms to fill the gaps:

Cyclones   **Typhoons**

Mexico   **Australia**

Country	Name used for tropical storms
	Hurricanes
Philippines	
	Cyclones
Bangladesh	

[2]

Total Marks _____ / 9

Tropical Storms – Case Study

1 Which of the following is most likely to have caused Typhoon Haiyan?

A   High pressure in the western Pacific

B   Warm surface water in the western Pacific

C   Cold surface water in the western Atlantic

D   Warm surface water in the western Atlantic

E   High pressure in the western Atlantic [1]

2 In the context of hazard management, what is 'inertia'? [2]

3 How can human factors contribute to a higher death toll when tropical storms hit poorer countries such as the Philippines? [3]

4 Draw lines to match the following responses to their types for a tropical storm in a country such as the Philippines:

Response   **Type of response**

Response	Type of response
Foreign investment in new infrastructure, e.g. more resilient bridges and railways	Immediate international response
Government declares state of emergency across the whole country	Long-term international response
Worldwide relief effort: aid valued at over $500 million	Long-term national response
Inertia tackled by offering incentives such as free bags of rice to encourage people to leave their homes	Immediate national response

[3]

Total Marks _____ / 9

Review Questions

Extreme Weather in the UK

1. Describe two of the main impacts of the UK drought of 2010–12. [2]

2. Give three strategies that were used to manage the UK drought of 2010–12. [3]

3. Explain how prevailing wind influences the UK climate. [3]

4. Draw lines to match the data below with each of the four climate zones shown on the map of the UK mainland.

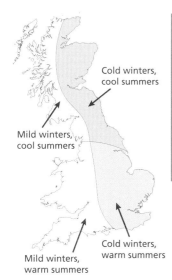

Cold winters, cool summers

Mild winters, cool summers

Cold winters, warm summers

Mild winters, warm summers

Place A	Annual rainfall 3000 mm	January average temperature 8°C	July average temperature 19°C
Place B	Annual rainfall 1000 mm	January average temperature 9°C	July average temperature 20°C
Place C	Annual rainfall 600 mm	January average temperature 3°C	July average temperature 15°C
Place D	Annual rainfall 660 mm	January average temperature 7°C	July average temperature 21°C

[3]

Total Marks _____ / 11

Climate Change

1. Which of the following best describes mitigation?

   A Making adjustments to reduce potential damage; limiting the impacts; taking advantage of new opportunities.

   B Cutting greenhouse gas emissions; reducing or eliminating long-term risk to human life and property; internationally agreed targets to reduce greenhouse gas emissions. [1]

2. What name is given to the changes in the Earth's orbit that can cause climate change? [1]

3. How can volcanic eruptions cause climate change? [2]

4. For these management of climate change strategies, decide which are adaptation (A) and which are mitigation (M).

Developing more drought-resistant crops or irrigation schemes.
Making buildings more energy efficient.
Higher flood defences along coasts and rivers.
Greater use of renewable resources.
Carbon capture and storage.
Planting more trees.

[6]

Total Marks _____ / 10

Ecosystems and Balance

1. Give an example of one small-scale ecosystem. [1]

2. Define the term 'ecosystem'. [2]

3. What is the difference between a food chain and a food web? [2]

4. Describe how the nutrient cycle works. [4]

Total Marks _____ / 9

Ecosystems and Global Atmospheric Circulation

1. In January, is high or low pressure found over central Asia? [1]

2. Define 'wind'. [2]

3. Name two features that can distort the behaviour of an air mass. [2]

4. Using examples, describe the global distribution of tropical rainforests. [4]

Total Marks _____ / 9

Practice Questions

Rainforests and Hot Deserts – Characteristics and Adaptations

1. What is the name of the soil found in a tropical rainforest? [1]

2. How do buttress roots help rainforest trees? [2]

3. Describe the climate in a tropical rainforest. [2]

4. Using examples, describe how animals have adapted to life in hot deserts. [3]

Total Marks _____ / 8

Opportunities, Threats and Management Strategies in the Amazon

1. Name two mineral deposits found under the Amazon rainforest. [2]

2. What is selective logging? [2]

3. Describe how tropical rainforests provide goods and services to humans. [4]

4. How can hydroelectric power schemes, like the Tucuruí project, create economic development? [2]

Total Marks _____ / 10

Opportunities, Threats and Management Strategies in Hot Deserts

1 What word describes roads, bridges, water and power lines? [1]

2 What do the letters HEP stand for? [1]

3 How has pastoral farming in Spain accelerated desertification? [2]

4 Describe how the problem of desertification can be tackled in the Mojave Desert. [4]

Total Marks _____ / 8

Polar and Tundra Environments

1 Where are tundra environments found? [1]

2 Name one benefit and one drawback of tar sand oil extraction. [2]

3 Describe the temperatures over the year in polar and tundra environments. [2]

4 Using examples, describe how plants have adapted to life in polar and tundra climates. [3]

Total Marks _____ / 8

Coasts 1: Processes

You must be able to:

- Describe the type and effects of waves
- Explain weathering, mass movement and erosion operating at the coast
- Describe how material is transported at the coast
- Explain why deposition takes place.

Coastal Processes and Landforms

- Factors affecting coastal processes and landforms include:
  - Waves: **fetch**; wave type and strength
  - Landscape processes: available material; type of erosion; amount of transport; deposition
  - The weather: rainfall; temperature change; winds; storm frequency
  - Geology: rock type; hardness; jointing; solubility
  - Biology: vegetation cover and plant root activity can damage or stabilise coastal areas; animal burrowing.

Waves

- Waves are caused by wind blowing over the ocean surface. Faster winds result in higher waves.
- Stronger waves have travelled furthest over open water (the fetch).
- Waves break in shallower water (at the plunge line) or against cliffs. The wave energy is pushed against the beach or cliff and transferred very quickly.
- Movement up the beach is **swash**. Return is **backwash**.
- **Plunging**: dig into the surface and remove sediment offshore.
- **Spilling**: long swash carrying material up a beach.
- **Surging**: base goes ahead and meets backwash of previous wave.
- **Constructive waves**: add material.
- **Destructive waves**: remove material; are more frequent in winter.

Rock Type and Structure

- More **resistant** (harder) rocks, such as granite and basalt, are less easily weathered than softer rocks, such as clay.
- Higher and steeper cliffs form on harder rocks (such as limestone and granite) but cliffs are lower and rounded on clay.
- Bands of hard and soft rock parallel to the sea make a **concordant** coastline. The harder rock protects the coast.
- Rock bands at right angles make a **discordant** coastline, with headlands and bays (see page 46).

Weathering and Mass Movement

- Weathering is the action of the atmosphere on rock.
- Rock is reduced in size (mechanical) and may be changed in composition (chemical).

Plunging Wave

Spilling Wave

Surging Wave

Constructive Wave

Wave breaks here

Swash

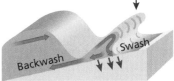

Large amount of water loss into the beach

Destructive Wave

Wave breaks here

Swash

Backwash

Small amount of water loss into the beach

Mechanical Weathering – Examples
• Frost action (water freezes in cracks, expands and causes rocks to break up).
• Salt crystal growth (salt from seawater dries in cracks and expands each time there is wetting and drying, causing rocks to break up).
• Plant roots, birds and animals burrow between joints, bedding planes and in cracks.

Chemical Weathering – Examples
• Rock **decomposes** and forms new materials.
• **Solution** creates holes and pitted surfaces; important on limestones.
• Acids from algae and seaweed attack rocks.

- **Mass movement** is the downslope movement of material under the influence of gravity. The material is then more easily eroded by waves.

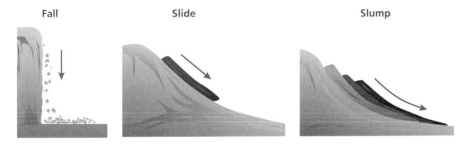

Fall   Slide   Slump

Erosion by Waves

Hydraulic Action

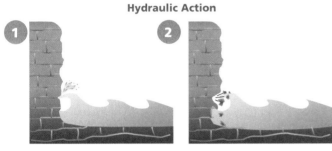

Abrasion

Attrition

Breaking waves force air into cracks in a cliff, exerting great pressure.

Sand, pebbles and boulders carried by waves are hurled against cliffs, weakening and breaking the rock. Very clear on cliffs with alternating harder and softer rocks.

Rock pieces detached from cliffs get smaller and rounder as they crash into each other moving up and down the beach.

Transport and Deposition

- Waves move material up and down a beach by **traction** (dragging along the floor) and **suspension** (carried in water).
- Backwash brings finer material back towards the sea, contributing to surface sorting by size.
- Wind makes sand grains bounce or jump (**saltation**).
- **Longshore drift**: material moves along the beach.
- Waves deposit material when they slow, e.g. in shallower water, after breaking.
- If more sediment collects than is removed, landforms are made.

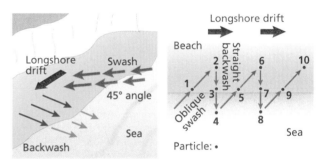

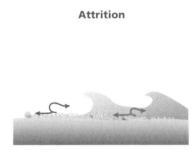

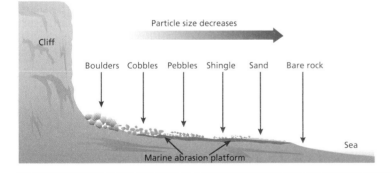

Particle size decreases

Cliff

Boulders Cobbles Pebbles Shingle Sand Bare rock

Marine abrasion platform

Sea

Key Words

fetch	solution
swash	mass
backwash	movement
plunging	hydraulic
spilling	action
surging	abrasion
constructive	attrition
wave	traction
destructive	saltation
wave	longshore
resistant	drift
decompose	

Coasts 2: Landforms

You must be able to:

- Describe headlands and bays and explain their relationship
- Explain different cliff shapes
- Understand the effects of rock type, structure, weathering and erosion on cliffs
- Describe the formation of erosional landforms: wave cut platforms, caves, arches, stacks and stumps
- Describe the formation of depositional landforms: beaches, sand dunes, spits and bars.

Headlands, Bays and Cliffs

- Wave action exploits weaknesses, e.g. faults or less resistant bands of rock on discordant coasts.
- There is deposition in bays.
- More resistant rock sticks out as headlands.
- Headlands shelter bays so material is deposited.
- Waves attack and erode headlands.
- Over time, coastlines get straighter.
- Cliffs are formed by weathering, mass movement and erosion.
- Rock type, rates of weathering and erosion are factors in the formation of different cliff shapes.

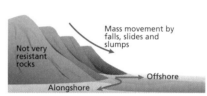

Harder rock
Headland
Beach
Softer rock
Bay
Harder rock
Headland

Slow Weathering

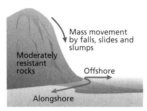

More resistant rocks
Mass movement mainly falls
Removal offshore
Alongshore

Sea more effective than weathering. Erosion and transport remove more material than inputs over time

Moderate Weathering

Moderately resistant rocks
Mass movement by falls, slides and slumps
Offshore
Alongshore

Balanced: Weathering and erosion plus transport even out over time

Rapid Weathering

Not very resistant rocks
Mass movement by falls, slides and slumps
Offshore
Alongshore

Weathering more effective than erosion and transport

Wave Cut Platforms, Caves, Arches, Stacks

- **Wave cut platforms** are gently sloping rock surfaces between the cliff base and the sea.
- Strong wave attacks create a **notch** at the cliff foot.
- Continued erosion causes cliff collapse, leaving a rock platform.
- There is usually a series of shallow steps that collect water. They are often covered with **debris**. They absorb wave energy so cliff erosion may cease.

Wave Cut Platform

Original land surface
Cliff
High tide
Low tide
Wave cut platform

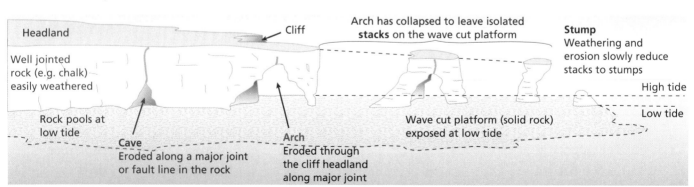

Headland

Well jointed rock (e.g. chalk) easily weathered

Rock pools at low tide

Cave
Eroded along a major joint or fault line in the rock

Cliff

Arch
Eroded through the cliff headland along major joint

Arch has collapsed to leave isolated **stacks** on the wave cut platform

Wave cut platform (solid rock) exposed at low tide

Stump
Weathering and erosion slowly reduce stacks to stumps

High tide

Low tide

Depositional Features

- **Beaches** form from loose material, e.g. sand, shingle, pebbles.
- Sandstones create boulders; chalk forms rounded pebbles; many rocks disintegrate to sand or mud-sized particles.
- Coarser material creates steeper beaches. Fine, sandy beaches can have a 3° slope; pebbles can make a 17° slope.
- Many beaches contain pebbles deposited by glaciers.
- If the coastline is parallel to approaching waves, material is deposited as wave energy is lost by friction near the land.
- Material is brought from **offshore** and alongshore and is removed by waves and longshore drift.
- The highest level of accumulation is on confined, bay head beaches where waves quickly lose energy.
- The beach height depends on the height of waves; the beach width on the amount of sediment arriving and being removed; the beach **gradient** on the wave steepness and particle size.
- Beach profiles change in winter and summer.
- **Sand dunes** are lines of sand hills.

Small bay head beach, Cornwall

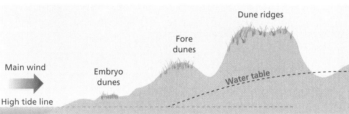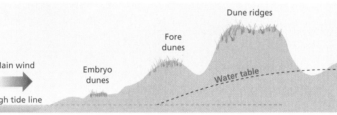

Waves rarely reach the back of a beach so the sand is dry and loose. Even between tides, it can dry enough to be bounced inland by an onshore wind.	Embryo dunes Sand grains move up the beach by saltation and get trapped by seaweed or debris. Plants, e.g. sea twitch or lyme grass, grow and bind the sand.	Fore dunes Tiny dunes join up to make yellow dunes. Made stable by marram grass.	Dune ridges Highest and biggest dunes. Look grey as plants die and decompose into the sand. Stop growing because as much sand is blown away as it arrives.

Dune ridges, Fore dunes, Embryo dunes, Main wind, Water table, High tide line

- Dunes form where there is a large store of dry sand, limited longshore drift to remove it, and consistent onshore winds.
- They are easily damaged by high winds and humans.
- **Spits** are long projections of deposited material where coasts change direction (often close to an **estuary**) and are joined to the mainland at one end.
- **Bars** are sand and/or shingle ridges parallel to the coast. They are submerged at high tide. Material is deposited offshore and builds up.
- Some bars are spits that grew across estuaries. Bars can stretch between headlands with a freshwater lagoon behind them.

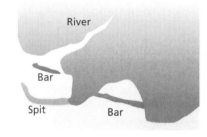

River, Bar, Spit, Bar

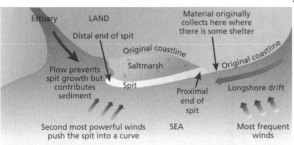

Estuary, LAND, Distal end of spit, Material originally collects here where there is some shelter, Original coastline, Saltmarsh, Original coastline, Flow prevents spit growth but contributes sediment, Spit, Proximal end of spit, Longshore drift, Second most powerful winds push the spit into a curve, SEA, Most frequent winds

1. Longshore drift brings sediment, which collects.
2. Finer sediment is collected in calmer water behind the spit.
3. Storm waves hurl pebbles on to the spit, helping growth and stabilisation.
4. River flow stops growth.
5. Waves make the end of the spit curve inwards.

Quick Test

1. What shape are cliffs that form on very resistant rock?
2. What is the seaward end of a spit called?
3. If a stack collapses, what feature results?
4. What is a bay?

Coasts 3: Management

You must be able to:

- Understand the threats to the coast
- Describe hard and soft engineering strategies
- Describe the management of an area of coastline.

Hard Engineering Strategies

- Hard engineering strategies stop wave and tide energy transferring to land.
- They are expensive to build and maintain.
- They are very effective at protecting the land behind but not further along the coast.
- They have a great impact on natural systems.
- Cliff strategies reduce wave energy and mass movement.
- Beach management strategies aim to increase beach size.
- **Sea walls** (a cliff strategy) are built parallel to the coast and reflect – not absorb – wave energy.
- There are different styles and shapes and they are expensive to build and maintain (thousands of pounds per metre).
- Most show serious damage after 30 years.
- **Rock armour** (a cliff strategy) is also known as rip-rap.
- Rocks are pushed into and on to the cliff face.
- The rocks' size and mass absorb wave energy. Gaps trap and slow down water, so materials are less likely to be removed.
- Rocks protect against damage from particles thrown up by waves.
- **Gabions** (a cliff strategy) are rocks held in wire mesh structures and are usually used in conjunction with other things, e.g. walls.
- They are quite ugly and relatively cheap, with a short life-span.
- They can restrict access to beach.
- **Groynes** (a beach management strategy) are built perpendicular (at right angles) to the coast and are made of concrete, steel or wood.
- They interfere with the movement of material carried by longshore drift.
- They create wider beaches.
- If they are too short, material is taken offshore.
- Material must be spread away from groynes back on to the beach to maintain their effectiveness.

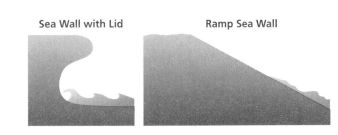

Sea Wall with Lid | Ramp Sea Wall

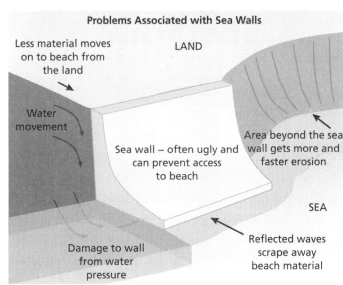

Problems Associated with Sea Walls

Less material moves on to beach from the land

LAND

Water movement

Sea wall – often ugly and can prevent access to beach

Area beyond the sea wall gets more and faster erosion

SEA

Damage to wall from water pressure

Reflected waves scrape away beach material

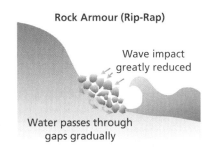

Rock Armour (Rip-Rap)

Wave impact greatly reduced

Water passes through gaps gradually

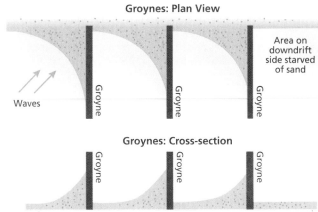

Groynes: Plan View

Area on downdrift side starved of sand

Waves

Groyne

Groynes: Cross-section

Groyne

- They do not restrict access to beach.
- The places beyond groynes (downdrift) are starved of sand.

Soft Engineering

- Soft engineering works with natural processes; however, it cannot stop erosion.
- **Beach nourishment** involves adding material to an existing beach; the material should match the existing beach in size and composition.
- The material can be taken from the offshore seabed but that disturbs wave patterns.
- Wave energy is reduced.
- Moderate cost: between £5000 and £200 000 per 100 m.
- The material can be imported, but this is more expensive.
- Beach nourishment increases the attractiveness of the beach.
- **Reprofiling** involves transferring material up or down the same beach, so material automatically 'matches' the existing beach.
- The new surface is about 1 m above the highest wave run-up.
- Reprofiling can repair dune damage (blow-outs) or rebuild longer lengths of beach and it speeds up dune recovery after storms.
- Material can be taken from behind groynes or breakwaters.
- Moderate initial costs but ongoing maintenance is required, which costs thousands of pounds per 100 m.
- **Dune regeneration** is an artificial version of a natural process.
- It is low cost (£200–£2000 per 100 m) but labour intensive.
- The dunes are easily damaged in storms.
- Fences are built at the seaward edge, and sand is transferred if necessary.
- Sand accumulates around the fences.
- **Marram grass** is planted above the highest wave reach to stabilise sand. **Lyme grass** and **sea couch** are planted below.

Managed Retreat

- Managed retreat is the removal of coastal protection, usually from areas that were claimed from the sea.
- It restores original sediment movements.
- Beaches recover and grow. **Saltmarsh** regrows.
- Costs are low: only the initial removal of old structures is needed.
- It eventually creates natural defences for the **landward** area.
- An example of managed retreat is Medmerry, West Sussex (2013).

Coastal Management at Mappleton	
Location	Holderness coast, East Yorkshire (3 km south of Hornsea)
Physical background	Boulder clay cliffs easily weathered and eroded. Fastest-eroding coast in Europe. Strong north-easterly winds from the North Sea. N–S longshore drift.
Human background	Small village and farmland on the B1242 (the only direct route along the coast), 50 m from the edge of the land.
Strategies	In 1991, Norwegian granite rock armour was built along the base of the cliff and two rock groynes installed.
Results	Erosion stopped between the groynes but increased to the south.
Other points	Cost £2 million; the improved tourist facilities meant this was mainly paid for with European Union funding. In 2015, cracks on the cliffs led to a land slip, making part of the beach unsafe.

Quick Test

1. To combat which process are groynes mainly built?
2. Why are there spaces in rock armour?
3. Which are more expensive: hard or soft engineering solutions?
4. What happens to most wave energy at a sea wall?

Rivers 1: Processes

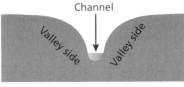

You must be able to:

- Explain the difference between a channel and a valley and name key parts of each
- Define river (fluvial) erosion, transportation and deposition
- Describe processes of erosion: hydraulic action; abrasion; solution; attrition
- Describe the ways in which material is carried by rivers
- Explain why deposition occurs
- Describe long and cross profiles of channels.

Channels, Valleys and River Systems

- A **channel** is the groove in which a river flows in a **valley** base.
- The most efficient channel is semi-circular and is twice as wide as it is deep.
- The **wetted perimeter** is the area touched by water.
- Water has to overcome friction to move.
- Changes to the channel affect the shape of the valley.
- The channel's **long profile** is from its **source** to its **mouth**.

A Channel's Shape is its Cross Profile

Channel

Valley side   Valley side

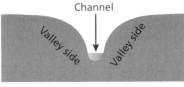

Valley side slopes

Channel

State of Water in a Channel

Bankfull: the river can do most work

Below bank

Materials on channel bed create friction

Parts of a River System

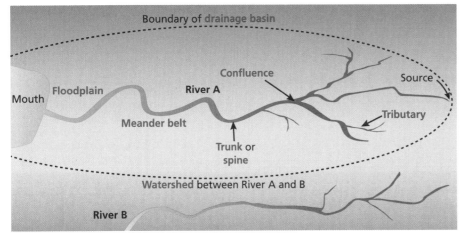

Boundary of drainage basin

Confluence

River A

Source

Floodplain

Mouth

Tributary

Meander belt

Trunk or spine

Watershed between River A and B

River B

River Erosion and Processes

- River (**fluvial**) erosion is the wearing away and removal of material from the channel. It needs more energy than for just moving water along.
- Material needs to be **entrained** and moved.
- Short periods of fast flow will cause more erosion than months of lower flows.
- Vertical erosion deepens the channel; lateral erosion widens the channel; **headward** erosion lengthens the channel.

Hydraulic Action Attacking the Side of a Channel

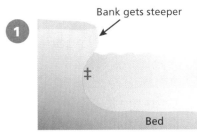

Bank gets steeper

1

‡

Bed

‡ Point of maximum speed and erosion

2

Bed

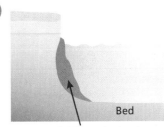

Material from the bank now available for river load

Hydraulic Action

- **Hydraulic action** is the action of moving water.
- Fast-flowing water will disturb loose material such as sand.
- **River banks** can be weakened and **undermined**, causing collapse.
- Hydraulic action is not effective on solid, hard rock.

Abrasion

- In **abrasion**, entrained material scrapes away the bed and banks.
- It can chip away at solid rock, creating small particles that can be carried along, and can smooth the bed and banks.
- Most **downcutting** of a channel is the result of abrasion.

Solution

- Dissolved material is carried away in **solution**. This is most effective on **limestones** but almost all rocks are partially soluble.

Attrition

- **Attrition** results from load particles crashing into each other.
- Particles being carried along get smaller and rounder.

Transportation and Deposition

- Material that has been entrained and then carried (**transported**) by a river is called its **load**.
- Faster flow enables more and bigger particles to be carried.
- Material carried in the body of the water is the **suspended load**.
- Material carried along the channel bed is the **bedload**.
- Bedload is moved by **traction** and **saltation**.
- Soluble material is carried in solution and is invisible.

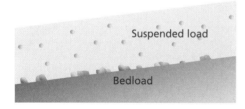

Suspended load

Bedload

- **Deposition** can occur at any point along the length of the river.
- It can be found in the same places as where erosion is taking place.
- Deposition occurs because there is not enough energy in the water to transport material.

Key Point

Coarser material like sand is easier to entrain but more difficult to carry; finer material like clay is harder to entrain but easier to carry.

Key Point

Attrition affects the load, not the channel, but it makes material easier to carry and to be used for abrasion.

Deposited material is visible in the channel

Very coarse bedload will only be moved when the river is high and moving rapidly

Key Words

channel	trunk	headward
valley	drainage basin	river bank
wetted	meander belt	undermined
perimeter	tributary	downcutting
source	floodplain	limestone
mouth	watershed	load
confluence	fluvial	suspended load
spine	entrained	bedload

Rivers 2: Landforms

You must be able to:

- Describe interlocking spurs, waterfalls, gorges, meanders, ox-bow lakes, floodplains, levées and estuaries, and explain their formation
- Understand the varying roles of erosion and deposition in making landforms
- Understand the relationship between landforms in a river landscape.

Interlocking Spurs

- Most river energy is used to overcome **friction** and move water.
- When flow is strong, small and highland streams mainly erode downwards by hydraulic action.
- Water flows around obstructions as it cannot wear through them.
- Land sticking into the route of the channel forms **interlocking spurs** that restrict views up or down a valley.

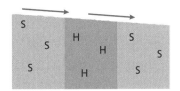
Interlocking spurs on a Pennine stream

Waterfall Associated with Horizontal Bands of Rock of Different Hardness

Waterfalls

- **Waterfalls** are a sudden steepening in the course of a river.
- They form where there are bands of rock with different hardnesses, **faulting**, at **plateau** edges and on hanging valleys.
- In waterfalls, water is not slowed by friction so deepens the pool at the front of the fall and the rock behind it is eroded by hydraulic action and abrasion.

Waterfall cuts back   Gorge

Plunge pool

Waterfall Associated with a Fault

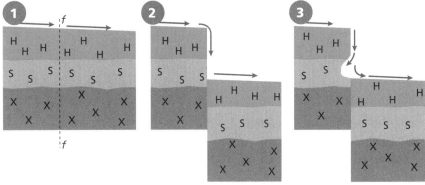

Waterfall Associated with Vertical Bands of Rock of Different Resistance

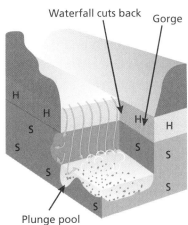

Gorges

- **Gorges** are steep-sided sections of a valley downstream from a waterfall. They can be formed in a number of ways:
  - Waterfalls eroding backwards into the rock as at the American Falls on the Niagara River.
  - Cavern collapse in limestone country as at Cheddar Gorge, Somerset.
  - Post-glacial floodwaters carving out new valleys with rapid downward erosion as at the Avon Gorge near Bristol.
  - In dry lands, erosion downwards rather than sideways creating canyons, as at the Grand Canyon in Arizona (although here the land has also been uplifted as the River Colorado has continued to cut down).

> **Key Point**
>
> Some features are a result of both erosion and deposition.

Meanders and Ox-bow Lakes

- **Meanders** are well-developed river bends. The most efficient route in a channel is the **thalweg**, which is **sinuous**.
- Water travels at different speeds.
- Faster water hits the outer bend, abrading the bank and making steep river cliffs. Flow at the outer bend is **helicoidal**.
- At inner bends, water is slower and deposition takes place; a **slip-off slope** forms.
- Meanders have an **asymmetric cross-section**.
- Over time, meanders move sideways and downstream.
- **Ox-bow lakes** are cut-off meanders.
- Meanders become increasingly developed, forming a narrow neck. At times of very high water flow, the meander neck is weakened, eventually breaking through.
- Water flows through both the meander and the straightened part but deposition seals off the meander, leaving an ox-bow lake.

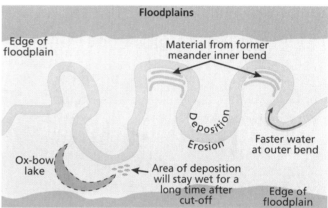

Helicoidal Flow – Cross-section

The water flows downstream from outer to the next inner bend

Water level · River cliff · Concave outer bend · Convex inner bend · Slip-off slope

Floodplains

- Floodplains form at the side of a river channel.
- They are made from **alluvium** – material carried and deposited by rivers. Some material may be from rivers overflowing their banks.
- The water not held in a channel spreads and slows, so even very fine material is deposited. Coarser material comes from the inner bends of meanders as they move sideways and downstream.

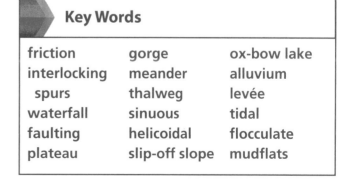

Floodplains

Edge of floodplain · Material from former meander inner bend · Deposition · Erosion · Faster water at outer bend · Ox-bow lake · Area of deposition will stay wet for a long time after cut-off · Edge of floodplain

Levées

- **Levées** are raised areas of coarser material, beside a river channel.
- At times of flood, water leaves the channel, suddenly slows and deposits material. The river bank is raised.

Levées

Edge of floodplain · Levée · Levée · Channel · Edge of floodplain

Estuaries

- **Estuaries** are **tidal** river mouths, found on rivers with wide mouths.
- River sediment **flocculates** (clumps together) when freshwater meets saltwater, and sinks. **Mudflats** form.
- Estuary material is generally finer towards inner areas and coarser at outer edges.

> **Key Point**
>
> Familiarise yourself with the landforms evident on the long profile of a UK river (e.g. the River Tees).

Quick Test

1. What is the steep-sided feature found in front of a waterfall?
2. What is the inner bend of a meander called?
3. What is most energy in a river used for?
4. What is alluvium?

Key Words

friction	gorge	ox-bow lake
interlocking spurs	meander	alluvium
waterfall	thalweg	levée
faulting	sinuous	tidal
plateau	helicoidal	flocculate
	slip-off slope	mudflats

Rivers 3: Flooding and Management

You must be able to:

- Explain how water gets into a channel
- Understand why channels fill more quickly or slowly
- Extract information from hydrographs
- Understand why floods occur
- Explain how floods can be prevented or their impacts reduced.

How Water Gets into a Channel

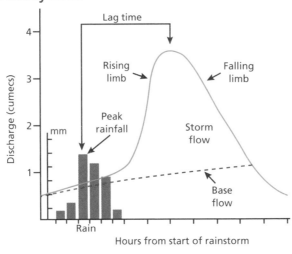

	Where from	Quick or slow?
Channel catch	Rain going directly in	Quick
Overland flow	Valley side surfaces	Quick
Throughflow	Soil under valley sides	Slow
Groundwater flow	Rocks below the river	Very slow

- Channels fill quickly if valley sides are **steep** (relief); if valley sides are impermeable (**geology**); if there is nothing to slow the water, e.g. trees (land use); if soils are already **saturated** (existing conditions).

Discharge and Hydrographs

- Discharge is the amount of water in a channel that passes a place at one time. It is measured in **cumecs** (cubic metres per second) and is calculated from the cross-sectional area of the water and its velocity.
- **Hydrographs** show river discharge over time.
- Steep rising and falling limbs mean discharge increases then decreases quickly – a **flashy** river.
- Gentle rising and falling limbs mean discharge changes slowly – a **subdued** river.
- **Lag time** is between peak rainfall and peak discharge.

KEY
...... Wetted perimeter   w   width
d   depth   v   velocity

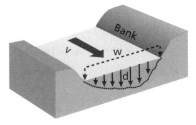

Cross-sectional area = w × d
Discharge = w × d × v

Contrasting Hydrographs

River A: flashy
River B: subdued

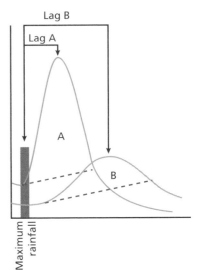

Floods

- Many British rivers overbank and **flood** once every 2–3 years.
- Channels are adjusted to 'usual' conditions.
- Water level rise can be dramatic, e.g. the River Valency in Cornwall rose 2 m in just under an hour on August 16, 2004.
- Causes of floods include: sudden, heavy storms; long periods of continuous rain; rapid melting snow and ice; high tides; increasing urbanisation; front lawns being replaced by tarmac for parking.

Preventing Floods

- The aim is to slow down water getting into the channel (abatement) and/or speed water through the channel.
- **Hard engineering:**
  - Artificial structures
  - Expensive
  - Intrusive on the landscape
  - Slow to build but have an immediate effect
- **Soft engineering:**
  - Working with the river landscape
  - Cheap; often zero cost
  - Can be slow to take effect
  - In tune with the landscape

York – A Complex Situation

- Low annual rainfall: 600 mm per year.
- Low-lying: 15 m above sea level.
- The rivers Ouse and Foss flow through the city. The rivers Swale, Ure and Nidd flow from the Yorkshire Dales into the Ouse.
- Changed land use, improved pasture and more building in the Yorkshire Dales speeds water into rivers.
- The North York Moors is overgrazed, has **peat** eroded by walkers and has lost forests, so water flows faster into the city.
- York is full of impermeable surfaces and has seen more and more house building.

York – Flood Defences

- Earth embankments around properties at the city edge.
- 'Washlands': store water from the Ouse when it rises.
- Concrete embankments and flood walls along river sides.
- Foss barrier: lowered to keep Foss water back when Ouse is high.
- The sunken gate gets raised to link city walls with flood barrier.
- Businesses and houses clear cellars to hold flood waters.

Hard Engineering Strategies
Straightening channels increases water velocity.
Dredging increases channel size but reduces speed, so more deposition.
Retention ponds allow water to go overbank into specific places.
Floodways alongside rivers take flood waters from the channel.
Dams regulate flow.
Embankments raise channel sides but move the danger downstream.

Soft Engineering Strategies
Trees **intercept** rainfall and take up water.
Terracing valley sides slows overland flow.
Removing boulders speeds up flow and are then used to raise and strengthen banks.
Restricting building on floodplains reduces impermeable surfaces.
Flood warnings, risk maps, preparation advice from the Environment Agency.
Earth embankments use local materials and blend into the landscape but can get waterlogged or destabilised by plants and animals.
River restoration puts river back to its natural state.

Key Words

- channel catch
- overland flow
- throughflow
- groundwater flow
- steep
- geology
- saturated
- cumecs
- hydrograph
- flashy
- subdued
- lag time
- flood
- intercept
- dredging
- terracing
- embankment
- peat

Quick Test

1. Which measurements are needed to calculate discharge?
2. What is lag time?
3. Give an example of an impermeable surface.
4. What is throughflow?

Glaciation 1: Processes

Quick Recall Quiz

You must be able to:

- Describe the distribution of UK upland and lowland areas
- Name and locate major UK river basins
- Define and explain freeze-thaw weathering
- Define glacial erosion, transportation and deposition
- Describe and explain abrasion and plucking (quarrying)
- Name and describe material transported by glaciers
- Explain why and how material is deposited by glaciers.

UK Landscape and Glaciation

- Ice sheets covered most of the UK for some of the last ice age.
- **Glaciers** were active in mountain areas, even when ice sheets were smaller.

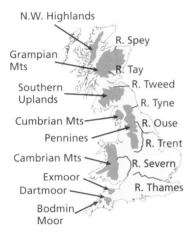

N.W. Highlands
R. Spey
Grampian Mts
R. Tay
Southern Uplands
R. Tweed
R. Tyne
Cumbrian Mts
R. Ouse
Pennines
R. Trent
Cambrian Mts
R. Severn
Exmoor
Dartmoor
R. Thames
Bodmin Moor

Maximum limit of ice sheet

> **Key Point**
>
> Most upland in the UK is in the north and west, on more resistant rock.

Frost-shattered rocks (scree) on Great Gable in the Lake District

Freeze-Thaw Action

- **Freeze-thaw** is a **weathering** process taking place above the ice.
- Temperatures need to go both above and below 0°C.
- Water needs to be present.
- Rocks need to have cracks or other weaknesses.
- The more enclosed the space in which water is trapped, the more effective the process.
- Rock is shattered into **angular** fragments.

Glacial Erosion

- Glacial **erosion** is the wearing away of the landscape, which can only happen when the ice is moving.
- Ice is very effective at altering the landscape, especially if it is laden with **debris**. The thicker the ice, the more erosion can occur.
- Erosion is evident in upland and lowland areas.

> **Key Point**
>
> Debris at the base of ice needs to be constantly renewed for erosion to be efficient.

Abrasion

- Rock fragments at the sides and base of a glacier cannot easily be pressed into the ice. If in contact with rock, such fragments will scratch the surface of the solid rock – this is **abrasion**. However, the debris needs to be harder than the rock being passed over.
- Scratches are created on the solid rock.

Abrasion
Particles at the base of the ice scratch the bedrock

Ice

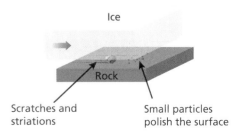

Rock

Scratches and striations

Small particles polish the surface

Plucking (Quarrying)

- For **plucking** to occur, the rock needs to have been weakened previously, perhaps by weathering.
- Moving ice melts when it meets an obstacle and the **meltwater** flows into cracks in the rock. As the glacier moves, the meltwater refreezes, taking the rock fragment with it.
- Ice is not strong enough to 'rip' solid rock from the landscape.
- Larger fragments can be removed by plucking than abrasion.

Rock surface on Snowdon showing evidence of abrasion and plucking

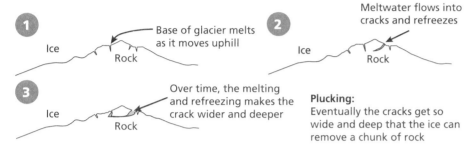

1 Base of glacier melts as it moves uphill
Ice
Rock

2 Meltwater flows into cracks and refreezes
Ice
Rock

3 Over time, the melting and refreezing makes the crack wider and deeper
Ice
Rock

Plucking:
Eventually the cracks get so wide and deep that the ice can remove a chunk of rock

Key Point
Erosion and deposition can occur at the same time.

Glacial Transportation and Deposition

- Debris **transported** by a glacier is called **moraine** and it can vary in size from tiny particles to huge boulders.
- The debris location gives the identity of the moraine:
  - in front of the ice: **terminal** moraine
  - under the ice: **sub-glacial** moraine
  - inside the ice: **en-glacial** moraine
  - on top of the ice: **supra-glacial** moraine
  - at the side of the ice: **lateral** moraine
  - where two glaciers meet, their lateral moraines may form a **medial** moraine.
- Material moves up and down within the ice.
- **Deposition** results from a reduction in the size and energy of a glacier, happening most rapidly at the front and edges of the glacier. Deposition is most evident in lowland areas.
- Deposits may be in a particular form or just a blanket of material called **till**.
- Deposits can be carried away and redeposited as **outwash** features.

Glacier retreating, i.e. forward movement of ice less than the rate of melting

Glacier snout

Lateral moraine

Valley side

Terminal moraine

Meltwater stream

Ground moraine (till) spread over the old landscape

Drumlins

Solid rocks

Erratic

Quick Test

1. Why is freeze-thaw action not erosion?
2. Which erosion process removes the biggest particles: abrasion or plucking?
3. Which is more likely to show evidence of both erosion and deposition: highland or lowland?
4. How would you recognise a piece of rock that resulted from freeze-thaw action?

Key Words	
glacier	moraine
freeze-thaw	terminal
weathering	sub-glacial
angular	en-glacial
erosion	supra-glacial
debris	lateral
abrasion	medial
plucking	deposition
meltwater	till
transportation	

Glaciation 2: Landscape

Quick Recall Quiz

You must be able to:

- Recognise and describe features of glacial erosion, including corries, arêtes, pyramidal peaks, glacial troughs (U-shaped valleys), truncated spurs and hanging valleys
- Explain the role of weathering and erosion processes in their formation
- Recognise and describe features of glacial deposition and explain distinctive forms
- Recognise the spatial and formation relationships between landforms.

Corries, Arêtes and Pyramidal Peaks

- High above the valleys, steep-sided, semi-circular rock basins form, called **corries**.
- Freeze-thaw action above the ice creates steep, high, back walls. Side walls are usually lower and less steep.
- At the base of the ice, abrasion deepens the basin and at the back of the ice, plucking makes the headwall steeper.
- Ice rotates through the basin to move downstream, causing over-deepening.
- Back-to-back corries make a knife-edged rock ridge (an **arête**).
- On large mountains, arêtes on three or more sides create a **pyramidal peak**.

U-Shaped Valleys and Troughs

- **Glacial troughs (U-shaped valleys)** are flat-floored and steep-sided with sudden closure at the upstream end **(trough end)** and are usually in former river valleys.
- Ice fills the valley, not just the river channel, to great thickness.
- Ice moving downslope erodes the base and sides of the valley, straightening its course. This is called 'all round erosion'.
- Freeze-thaw action above the ice brings debris on to the glacier surface, widening the valley.
- Abrasion and plucking occur where the ice touches the rock, scouring away material, including former interlocking spurs, to create **truncated spurs**.
- After the disappearance of the ice, the original valley floor is rocky and irregular. Infilling by material washed from glaciers and rivers creates a smooth, flat floor.
- **Ribbon lakes** form on the valley floor and are gradually filled in. Today, some but not all valleys have lakes.

Hanging Valleys

- Tributary valleys filled with smaller, less-powerful glaciers so were much less eroded. Their ice contributed to the main glacier. Former confluences are eroded away, leaving tributary valleys **'hanging'**.
- Ice from tributaries creates ice-falls into the main valley, which may become waterfalls.

> ### Key Point
>
> All erosion landforms result from both weathering and erosion.

Formation of Glacial Troughs

Water erodes the channel mainly downwards

Ice erodes "all round"

The valley becomes the channel for ice

River flows between interlocking spurs

The ice carves away the land where the spurs interlocked

Landforms of Glacial Deposition

- An **erratic** is a glacially transported block left behind in an area of different rock type.
- **Drumlins** are rounded, elongated, asymmetrical mounds of moraine shaped by moving ice. The 'blunt' end faces the direction of ice approach. The ice tapered the tail as it flowed over. Drumlins occur in groups.

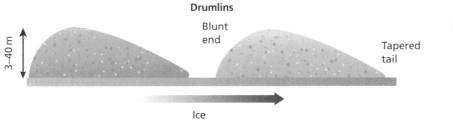

Drumlins

Blunt end

Tapered tail

3–40 m

Ice

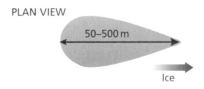

PLAN VIEW

50–500 m

Ice

- **Ground moraine** or **till** is a mixture of rock particles in clay smeared on to the surface and is often very thick.
- Lateral moraines are ridges of debris along the side of the valley.
- Terminal moraines are crescent-shaped ridges, tens of metres high in the UK, made of unsorted material and stretching across a glaciated valley marking the furthest limit of ice.
- Moraines are not solid so are easily changed by meltwater, weathering and erosion by rivers.

Glaciated Upland Landscape

- Langdale Valley in the Lake District shows a range of ice-formed features and post-glacial change.

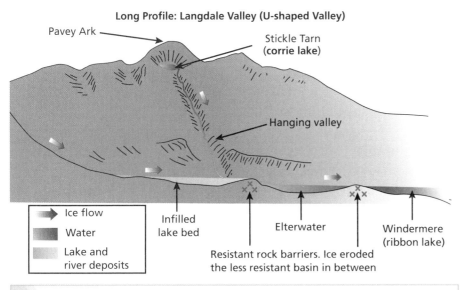

Long Profile: Langdale Valley (U-shaped Valley)

Pavey Ark

Stickle Tarn (**corrie lake**)

Hanging valley

Ice flow

Water

Lake and river deposits

Infilled lake bed

Elterwater

Windermere (ribbon lake)

Resistant rock barriers. Ice eroded the less resistant basin in between

Quick Test

1. Where would an arête form?
2. Name two features evident along the sides of a trough.
3. What would a glaciated valley floor look like immediately after the ice has disappeared?
4. What is till?

Glaciation 3: Land Use and Issues

You must be able to:

- Understand how glaciated highlands provide opportunities for, and limits to, land uses
- Describe the range of economic activities found in glaciated highland areas: industry, farming and tourism
- Explain a range of potential conflicts and management strategies.

Landscape Characteristics

- Glaciated uplands are high with steep slopes and are often **impermeable**.
- Soils are not deep, and can be **waterlogged** on flatter areas or easily **eroded** from steep slopes.
- The uplands contain different lake types (such as ribbon lakes and corrie lakes). There are many rivers.
- The uplands receive high rainfall (in excess of 1500 mm per year).
- The high altitude means that there is a short **growing season**.
- The high **terrain** means exposure to winds.
- There are low sunshine totals of under 1000 hours per year.
- High rainfall in glaciated uplands, as well as steep slopes and impermeable rock, results in rapid stream flow.
- Old **industries** used the local water supply as a raw material and power source. Nowadays, hydroelectric power (HEP) can use the fast-flowing water to generate energy.
- Lakes offer a water supply but are also a potential **flood** danger.
- Resistant rocks, such as slate and granite, provide useful resources for buildings, roofs and roads.
- Metal ores are often found in glaciated highland areas and were extensively mined.
- Nowadays the extraction of minerals attracts opposition for **environmental** and **aesthetic** reasons.
- **Pastoral farming** is more important than **arable**.
- Hardy crops, for **fodder** or **silage**, are grown on lower, flatter ground.
- **Dairy** cattle are kept on lower ground and sheep on higher ground.
- Farmland is divided into areas according to the **relief**.

Land Use on a Lake District Farm

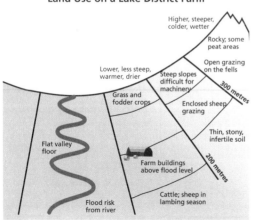

Land Use on a Scottish Upland Farm

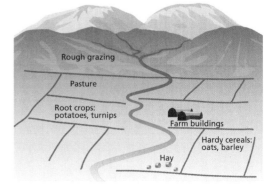

Tourism and Conflicts

- The scenery attracts visitors for a range of **tourism**, including both sporting and non-sporting activities.
- Farmland is often used for tourist activities.
- **Diversification** on farms is often related to tourist activities, e.g. the conversion of barns into accommodation.
- Old mine workings provide historical and tourist interest.
- Tourists are needed to maintain and improve the economy but can cause conflict.
- Areas such as the Lake District and Snowdonia are very popular.

Conflicts	Examples
People and the landscape	Footpath erosion; gullying; grass or vegetation loss; litter
Tourists and locals	Crowded streets; rise in property prices due to demand for holiday lets; changing services to cater for tourists
Tourists and farmers/ landowners	Damage to crops; gates left open; rubbish in watercourses
Traffic and the landscape	Increased impermeable surfaces affects hydrology; air **pollution** from exhausts; road-building can change slope angles; visual pollution from proliferation of signs.

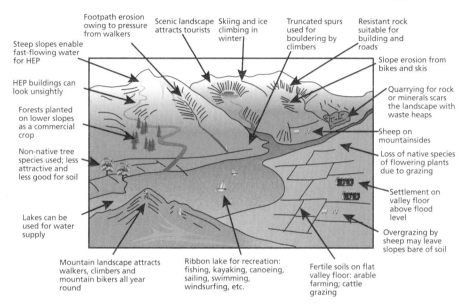

Footpath erosion owing to pressure from walkers

Scenic landscape attracts tourists

Skiing and ice climbing in winter

Truncated spurs used for bouldering by climbers

Resistant rock suitable for building and roads

Steep slopes enable fast-flowing water for HEP

HEP buildings can look unsightly

Forests planted on lower slopes as a commercial crop

Non-native tree species used; less attractive and less good for soil

Lakes can be used for water supply

Slope erosion from bikes and skis

Quarrying for rock or minerals scars the landscape with waste heaps

Sheep on mountainsides

Loss of native species of flowering plants due to grazing

Settlement on valley floor above flood level

Overgrazing by sheep may leave slopes bare of soil

Mountain landscape attracts walkers, climbers and mountain bikers all year round

Ribbon lake for recreation: fishing, kayaking, canoeing, sailing, swimming, windsurfing, etc.

Fertile soils on flat valley floor: arable farming; cattle grazing

Management Strategies

- Major policies, such as the Lake District National Park Local Plan 2020 to 2035, can help to manage the conflicting interests and plan for a sustainable future.
- Individual strategies to help mitigate conflicts can include:
  - footpath schemes (e.g. Fix the Fells in the Lake District)
  - house-building schemes aimed at locals
  - improved bus services and cycle paths
  - provision of more litter bins
  - information boards and exhibitions provided by national parks authorities.

Key Words

impermeable
waterlogged
erosion
growing season
terrain
climate
industry
flood
environmental
aesthetic
pastoral farming
arable
fodder
silage
dairy
relief
tourism
diversification
pollution

Quick Test

1. What is diversification?
2. Name one sporting activity associated with glaciated upland areas.
3. Give one non-recreational use (past or present) of the rivers or lakes.
4. What is the main type of pastoral farming found in glaciated uplands?

Review Questions

Ecosystems and Balance

1 Give an example of a global large-scale ecosystem. [1]

2 Define the term 'producer'. [2]

3 Define the term 'consumer'. [2]

4 Define the term 'decomposer'. [2]

Total Marks _____ / 7

Ecosystems and Global Atmospheric Circulation

1 Which line encircles the Earth and divides it into the Northern Hemisphere and the Southern Hemisphere? [1]

2 Which line encircles the Earth at roughly 23° **north** of the Equator? [1]

3 Which line encircles the Earth at roughly 23° **south** of the Equator? [1]

4 Using examples, describe the global distribution of the tundra ecosystem. [4]

Total Marks _____ / 7

Rainforests and Hot Deserts – Characteristics and Adaptations

1 In tropical rainforests, where are the majority of nutrients found? [1]

2 What word describes the artificial watering of crops? [1]

3 How have pitcher plants evolved to cope with the low nutrient soils in rainforests? [2]

4 Describe the climate in hot deserts. [2]

Total Marks _____ / 6

Opportunities, Threats and Management Strategies in the Amazon

1 What do the initials CITES stand for? [1]

2 What happens to the soil after vegetation cover is removed? [1]

3 What is the difference between commercial and subsistence farming? [2]

4 Using examples, describe how a named HEP scheme can have both benefits and drawbacks. [4]

Total Marks _____ / 8

Opportunities, Threats and Management Strategies in Hot Deserts

1 Name one HEP scheme in the Mojave Desert. [1]

2 The mining of which mineral has led to a reduction in underground aquifers? [1]

3 What are 'terraced' fields? [2]

4 Describe how the problem of desertification can be tackled in southern Spain. [4]

Total Marks _____ / 8

Polar and Tundra Environments

1 What word describes the characteristic landscape type in tundra regions? [1]

2 Name two international agreements that control the human use of Antarctica. [2]

3 Using examples, describe how animals have adapted to life in polar and tundra climates. [4]

4 Name three resources found in and around Antarctica. [3]

Total Marks _____ / 10

Practice Questions

Coasts 1: Processes

1 Explain why beach material gets smaller and rounder as you move from a cliff to the sea. [2]

2 Explain why different rock types produce different shapes of cliff. [4]

Total Marks .... / 6

Coasts 2: Landforms

1 Look at the photograph of part of the Shiant Islands in Scotland.

Feature A

a) What is feature A? [1]

b) With continued erosion, what feature might be formed in the same place? [1]

2 Why might a spit have a curved end? [2]

3 What conditions are best for the creation of sand dunes? [3]

Total Marks .... / 7

Coasts 3: Management

1 Explain **one** negative point about sea walls as coastal protection. [3]

2 Compare beach nourishment and beach reprofiling. [4]

Total Marks .... / 7

Rivers 1: Processes

1 On what type of rocks is solution most effective? [1]

2 Describe one way in which bedload can be moved. [3]

3 Explain the connection between entrainment and erosion. [3]

> **Total Marks** .... / 7

Rivers 2: Landforms

1 Look at the photograph of a river in Scotland.

   a) Name the features labelled A and B. [2]

   b) What feature is likely to form at C? [1]

   c) What part of the river is at location X? [1]

   d) What process is operating at location X? [1]

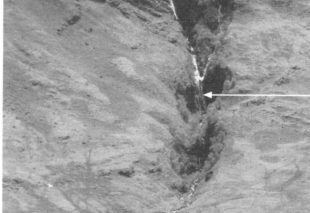

Location X

A

B

C

2 Draw the cross-section of a channel at a meander bend.
Add these labels: Inner bend; Outer bend; Deposition; Erosion. [4]

Total Marks _____ / 9

Rivers 3: Flooding and Management

1 Give two reasons why a river might be 'flashy'. [2]

2 Look at the storm hydrograph below.

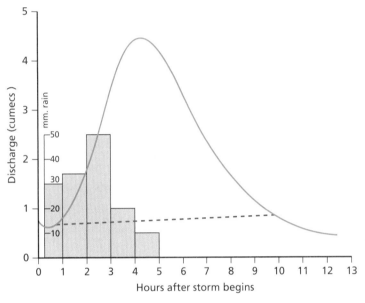

a) What was the lag time? [1]

b) What was the highest rainfall? [1]

c) What was the highest discharge? [1]

d) For how many hours was the river above normal flow? [1]

3 Describe one **slow** way in which rainwater can reach a channel. [4]

Total Marks _____ / 10

Glaciation 1: Processes

1. What is the material called that is pushed along by a glacier and deposited at its furthest point? [1]

2. Where does abrasion take place? [2]

3. On the diagram below, insert an arrow to indicate the direction of ice flow and label these in the correct places: Bedrock; Plucking; Debris; Plucked block; Abraded surface. [6]

Total Marks .... / 9

Glaciation 2: Landscape

1. Where is the trough end in a glaciated valley? [1]

2. Describe the appearance of the back wall of a corrie. [2]

3. Why might a glaciated valley be flat and quite smooth today but was rocky just after the ice disappeared? [4]

Total Marks .... / 7

Glaciation 3: Land Use and Issues

1. Why is flooding a potential risk in glaciated areas? [4]

2. How do glacial landforms favour recreation activities? [4]

Total Marks .... / 8

Urbanisation

You must be able to:

- Understand how rapidly the world's urban population is growing
- Explain the reasons why people migrate to urban areas.

Quick Recall Quiz

Urbanisation

- **Urbanisation** means an increase in the proportion of people living in urban (town and city) areas.
- In 1900, 10% of the world's population lived in urban areas. This had risen to 34% by 1960 and to 56% by 2020. The United Nations predicts that 68% of the world's population will be urban by 2050.
- Levels of urbanisation are increasing across the world, with the fastest rates of growth occurring in lower income countries (LICs).
- As countries industrialise, levels of urbanisation increase as people move into cities for jobs.
- Most higher income countries (HICs) industrialised over 100 years ago and now have large urban populations. LICs are still in the early stages of industrialisation and so their urban populations are still growing.
- As urban areas grow outwards, they encroach into rural (countryside) areas. This is known as **urban sprawl**.

World Megacities

- A **millionaire city** is one with a population of over 1 million people. There are currently 280 millionaire cities in the world, and the number is growing rapidly – mostly in LICs. It is predicted that there will be 663 millionaire cities by 2030, with 148 of those in China.
- A city with over 10 million people is known as a **megacity**. There were 42 in 2022, mainly in LICs and newly emerging economies (NEEs).
- Examples of megacities include Shanghai (China), Beijing (China), Delhi (India), Karachi (Pakistan), Istanbul (Turkey) and Lagos (Nigeria).
- LIC cities are growing so fast due to rural-to-urban migration and high natural increase in population (high birth rates and falling death rates).

Key Point

The majority of the world's population now lives in urban areas.

Key Point

Most of the world's largest cities are in LICs.

Rural to Urban Migration

- **Push factors** are those that drive people away from rural areas, e.g.:
  - unemployment
  - low wages
  - drought, famine and other natural disasters
  - farming is difficult and unprofitable
  - few job opportunities
  - lack of social amenities
  - isolation
  - civil war.
- **Pull factors** are those that attract people into cities, e.g.:
  - more job opportunities
  - higher wages
  - better schools and hospitals
  - better housing and services (like water, electricity and sewerage)
  - better social life
  - better transport and communications.

City life can attract people from rural areas

PULL FACTORS

Better education

More houses

Better healthcare

More wealth

Communications and power

Poverty

Not enough land

Natural disasters

Crop failures

FARMING VILLAGES

Remoteness

PUSH FACTORS

Key Point

LIC cities are growing rapidly, mainly due to rural to urban migration.

Quick Test

1. What is a millionaire city?
2. What percentage of the world's population was urban in 2020?
3. How many people are needed to qualify a city as a 'megacity'?

Key Words

urbanisation
millionaire city
megacity
push factors
pull factors

Quick Recall Quiz

Urban Issues and Challenges 1

You must be able to:

- Explain how urban growth creates opportunities and challenges within cities in LICs and NEEs
- Explain different ways conditions in shanty towns can be improved for the residents.

Case Study: City in a Newly Emerging Economy (NEE) – Rio de Janeiro, Brazil

- Population: 13.6 million (2022).
- 24% of population live in over 1000 **favelas**.
- Location: south-east coast.
- Reasons for growth: rural to urban migration and natural population increase.
- Former capital city of Brazil.
- Major tourist attractions: Copacabana Beach, statue of Christ the Redeemer, Sugar Loaf Mountain.
- Hosted football's World Cup in 2014 and the Olympic Games in 2016.

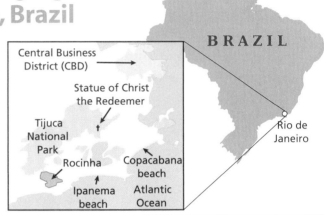

> ### Key Point
>
> Almost a quarter of Rio's population live in favelas.

Challenges in Rio

Social	**Migration:** Rapid growth in recent years because migration from rural areas to the city has put huge pressure on services and amenities. Migrants have been attracted to Rio due to job opportunities, higher wages and better services like schools and hospitals.
	Housing: Rio has a number of shanty town settlements (favelas) that house the many rural migrants who come to the city. They are unplanned and spontaneous – often growing up on poor quality land. Residents have no legal land ownership. Houses are built using cheap materials like wood or corrugated iron and many people are crammed into a small area, leading to overcrowding, with no clean running water or sewage disposal. There are no schools or hospitals, few job opportunities and high levels of disease and illness.
	Healthcare: Has been poor but more clinics have now been established, especially to improve maternity care and support for the elderly.
	Education: Schools suffer from low enrolments, and funding has been made available for higher pay and better training for teachers, as well as grants for poor families.
	Water supplies: A clean water supply is not available for 12% of the population. New treatment plants have been built and new pipelines laid.
	Energy: A shortage of electricity has meant frequent blackouts. New power lines have been constructed, along with a nuclear power station and a new HEP station.
	Crime: Street crime has been high, with powerful drug gangs controlling the favelas. The police have taken steps to reduce crime, including the 'Pacifying Units' that have reclaimed order in some of the favelas.
Economic	**Poverty:** There is a huge gap between rich and poor citizens in Rio. Some favelas have grown on hillsides right next to the affluent central business district.
	Employment: Unemployment remains high in the favelas (over 20%), where most people work in the informal economy. Poor transport systems make it difficult for favela dwellers to access other parts of the city.
Environmental	**Urban sprawl:** As the city continues to grow, it encroaches on surrounding rural areas.
	Pollution: Air pollution from heavy traffic and congestion is a major issue, as is pollution of the sea from sewage and industrial waste.
	Waste disposal: This is a particular problem in the favelas, many of which are inaccessible to collection vehicles.

Opportunities in Rio

Social	Cultural and ethnic diversity: Rio is home to a huge mix of different races, religions and cultures. The annual Rio Carnival is an international event.
	Education: Rio has a number of universities and is a centre for research and development in Brazil.
	Strength of community: There are many positive aspects to life in favelas, including the people who create their own economy, e.g. shops, restaurants and cottage industries (e.g. pottery); residents send money back home to families in rural villages; recycling commonly takes place, for example of waste and building materials.
	Transport: Economic development has led to improvements in roads and transport systems.
Economic	Industry: Industrial growth has boosted Rio's economy through steel making, port industries, oil refining, petrochemicals, manufacturing (including modern industries like computers and electronics) and a growing range of services such as banking and finance.
	Tourism: Rio is one of the most-visited cities in the Southern Hemisphere. Major attractions include Copacabana Beach, Ipanema Beach, the statue of Christ the Redeemer and Sugar Loaf Mountain.
Environmental	Beaches: The Atlantic beaches continue to attract tourists.
	Forests: The Tijuca National Park is one of the largest urban forests in the world.

Rocinha Shanty Town, Rio de Janeiro

- The rapid growth of Rio's population has led to a severe shortage of housing. The city has a number of shanty town settlements, such as Rocinha, that house the rural migrants who flock there.
- Rocinha is the largest shanty town in Rio, with a population of around 200 000 people. It is located in the south zone of the city on a steep hillside overlooking the city beaches.
- Not all the people in Rio are poor – many wealthy people live close to the central business district (CBD).

Improving the Quality of Life in Rocinha

- The Favela Bairro Project (a slum to neighbourhood project), which ran from 1995 to 2008, was set up to upgrade the favelas (as opposed to demolishing them) by providing pavements, electricity and sewerage systems.
- The project promoted self-help building schemes, where local residents were provided with materials such as concrete blocks and cement to construct permanent dwellings, often with three or four floors and with water and sanitation.
- Residents were given legal rights of ownership or low rents on properties. Improved transport systems gave them better access to work in the city.
- Businesses like shops and restaurants were encouraged.
- Law and order improved through 'pacification' programmes.
- After the area was made safe for visitors, tourism flourished.

Key Point

Rapid urban growth of Rio has presented many social, economic and environmental challenges.

View over Rocinha favela

Key Point

Conditions in favelas can be improved with government support and self-help projects.

Quick Test

1. What percentage of Rio de Janeiro's population live in favelas?
2. Explain three ways in which living conditions in Rocinha favela have been improved.
3. Explain how self-help schemes have improved conditions in Rocinha favela.

Key Words

favela

Urban Issues and Challenges 2

You must be able to:

- Explain how urban changes in cities in the UK lead to a variety of social, economic and environmental opportunities and challenges
- Explain different ways inner city areas can be regenerated.

Case Study: UK City – London

- Population: 9.3 million (2021).
- Leading global city for industry, education and finance.
- A world cultural capital.
- The most-visited city in the world.
- Diverse range of peoples and cultures.
- Hosted the Olympic Games in 2012.
- During the 19th century, the port of London was the busiest in the world. By 1980, ships had become too large to sail up the River Thames and the docks close to the centre became derelict.

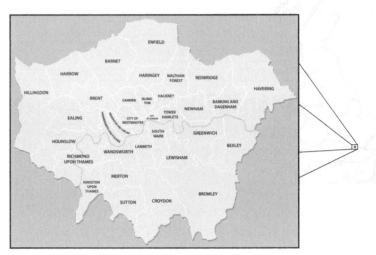

Challenges in London

Social	
	Housing inequalities: House prices in London are greater than anywhere else in the UK. Within the city, there is not enough good quality and affordable housing. Overcrowding has risen (particularly in the private rental sector) and is considerably higher than the rest of the country.
	Education: Overall attainment in London schools matches the rest of the UK, although there is considerable variation across boroughs in the city.
	Health: People with poorer English language skills have found it difficult to access healthcare facilities.
	Urban sprawl: As London has expanded, new buildings have been constructed on greenfield sites around the city.
	Migration: Job opportunities, particularly in finance and 'knowledge-based' industries, have attracted people to London from elsewhere in the UK and the rest of the world. It is estimated that up to one-third of all international migration into the UK is to London.
Economic	**Industry:** As the docks closed, many manufacturing industries were lost.
	Pay inequality: London has the most unequal pay distribution of any part of the UK, largely due to the high wages paid at the top end of the scale. 21% of people living in London are paid below the London living wage.
	Unemployment: The closure of factories led to high unemployment in parts of inner London.
Environmental	**Dereliction:** As manufacturing industries have declined (particularly in the eastern parts of the inner city and along the banks of the River Thames), much land has been left in a state of dereliction.
	Urban sprawl: This has increased pressure on land use in the rural–urban fringe and led to the growth of commuter settlements.
	Waste disposal: A large urban population produces a lot of household and commercial waste for disposal.
	Pollution: Atmospheric pollution from industry and vehicles.

Opportunities in London

Social	Diversity: Ethnic and cultural diversity allows people to experience different foods, music and religions.
	Entertainment and culture: London is an international centre for sporting events, theatres, cinemas, museums and art galleries.
Economic	Industry: The number of job opportunities in financial services and knowledge-based industries has increased. In addition, the redevelopment of London's docklands and the more recent redevelopment of the Olympic site in East London have increased the number and variety of jobs available in London.
	Transport: A new high-speed train link (HS2) from London to Birmingham and eventually on to Manchester is planned. The Crossrail project from Paddington station to Reading in the west and Abbey Wood to the east of the city becomes fully operational in 2023. Plans have been approved to build a third runway to expand Heathrow Airport.
	Tourism: London is the most-visited world city and attracts tourists with its historic buildings, monuments, sports events, cultural events and entertainment.
Environmental	Reuse of industrial land: Numerous brownfield sites have become available for development.
	Transport: Improvements to transport systems aim to reduce carbon dioxide emissions by 60% by 2025. This includes wider use of diesel-electric hybrid buses, and the trialling of hydrogen fuel cell buses, and combined diesel and biofuel buses.
	Regeneration: Much of the industrial Thames riverside has been regenerated or re-imaged. The docks were regenerated in the 1980/90s into offices on Canary Wharf and new housing throughout the area. Specific sites have also been re-imaged: the Bankside Power Station is now the Tate Modern and other buildings have been converted into living accommodation (e.g. Battersea Power Station).

Regeneration of the Lower Lea Valley

- The lower Lea Valley was a heavily polluted, brownfield area and was the site of various toxic industries, some now derelict, and the extensive marshalling yards at Temple Mills.
- It was decided when London bid in 2005 for the 2012 Olympics that this 560 ha area was going to be used for the various stadia, athletes' village and all the other requirements of a modern summer Olympics.
- Links were made to the underground system, the Docklands Light Railway (DLR) and mainline rail services at the new railway and bus interchange.
- The Olympics and Paralympics cost nearly £9 bn in total.
- The plan was to create 40 000 local jobs by 2025 and 24 000 new homes by 2031.
- Since 2012, the post-Olympics landscape has been transformed with more than 35 km of pathways and cycleways, 6.5 km of waterways, 4300 trees, playgrounds and a park hosting year-round events and sporting activities.
- The Olympic Stadium became home to West Ham United FC.
- Much of the affordable housing still remains out of reach for some people and gentrification has pushed up prices locally.
- Despite all the positive benefits, the negative effects mean that the regeneration of this area remains controversial.

Stratford's Olympic Stadium as seen from the River Lea. The ArcelorMittal Orbit, an observation tower and Britain's largest piece of public art, stands alongside.

Quick Test

1. List three economic challenges faced by London.
2. Name two projects aimed at improving transport in London.
3. How was the Olympic Stadium used after the 2012 Games?

Key Words

brownfield site
regeneration

Urban Issues and Challenges 3

You must be able to:

- Understand why urban areas need to be developed for the future in a sustainable way
- Explain how sustainable living involves a number of changes to current urban lifestyles.

Sustainable Urban Living

- City inputs include food and water, fuels and energy, building materials and consumer goods.
- City outputs include sewage, exhaust gases, household waste, industrial waste and building waste.
- In the UK, 90% of people live in urban areas.
- Cities use huge volumes of resources and produce vast amounts of waste.
- However, urban areas can be developed in a **sustainable** way.
- Sustainable living means allowing people to meet their needs today without harming the prospect of people in the future to meet their needs.
- Sustainable urban living needs to consider ways for people in towns and cities to adapt to climate change and use strategies to mitigate it.

Features of Sustainable Urban Living

Housing
- Affordable prices and low rents.
- Shared housing.
- Passive housing – kept warm using heat from people, pets and natural lighting.

Resource Management
- The aim is to be **carbon neutral** – for a town to produce as much energy as it consumes.
- Low energy use, e.g. solar heating, efficient insulation, triple glazing, water and electricity meters, low-flow taps, etc.
- Efficient recycling.
- Reduction of household and industrial waste.
- Composting of food and green waste, which counters global warming by reducing methane emissions.
- 'Greywater' and rainwater are collected for domestic use.
- Sedum moss roofing is used to reduce impermeable surfaces, increase lag times and reduce flooding.

> **Key Point**
>
> Many current trends in urban living are not sustainable. Sustainable urban living includes methods of adaptation to, and mitigation of, climate change.

> **Key Point**
>
> Sustainable modern cities are self-contained, self-supporting and help to manage climate change.

Urban Transport Strategies to Reduce Traffic Congestion

- Car-sharing schemes.
- Vehicle-restricted areas to reduce congestion and pollution.
- Extension of congestion charges.
- Public transport (including buses, trains and trams) organised into a diverse and efficient integrated rapid transit system.
- Alternative, cleaner fuel sources, e.g. hydrogen and electric. New electric buses planned for Bristol are designed to charge with power when parked over induction plates. These could eventually be installed along the bus routes to allow the vehicles to charge on the move.
- Bus lanes.
- Park-and-ride systems.
- Increased bicycle use, e.g. a system of public cycle hire was introduced in London in 2010 and became known as 'Boris' Bikes', after Boris Johnson, the London Mayor from 2008 to 2016. By 2021, 11 500 bicycles had been made available for hire from 750 docking stations distributed around the city.
- Pedestrian walkways and cycle paths, e.g. London's cycle 'super-highways'.

Services and Employment

- Daily needs, such as shops and schools, should be within walking distance for the population.
- There should be a range of job opportunities for residents.

Environment

- 'Urban greening' – a target of 40% green space, parkland and trees.
- Urban forests, e.g. Adelaide in Australia.
- Urban agriculture – vertical farming, allotments and rooftop gardens with the aim of reducing 'food miles'.
- Use of brownfield sites (previously used for industry) for development.
- A green belt of natural space should surround any settlement.

Measuring Development and Quality of Life

You must be able to:

- Identify ways by which development and quality of life can be measured and recognise some of the limitations of these measures
- Classify countries by levels of economic development and quality of life
- Make links between stages of the Demographic Transition Model and level of development.

Measuring Development

- **Development** refers to the progress of a country in terms of economic growth, human welfare and the use of technology.
- **Quality of life** refers to the wide range of human needs that should be met alongside income growth, such as access to housing, education and health, nutrition, security, happiness, etc.
- There is more to development and quality of life than just wealth.
- Geographers use a variety of data (or indicators) to decide how developed a country is. These indicators can cover social, economic or environmental factors such as those shown to the right.
- One single factor does not always tell a full and accurate story about a country. It has been popular in recent times for geographers to use the **Human Development Index** (HDI), which combines life expectancy, literacy and income to measure development.

Indicators used to determine a country's development

- GNI (gross national income) and GDP (gross domestic product), often expressed in US dollars
- Life expectancy (expected number of years of life)
- Literacy (percentage of population that can read and write)
- Birth rate (number of babies born per 1000 people per year)
- Death rate (number of deaths per 1000 people per year)
- People per doctor (population total divided by number of available doctors)
- Calorie intake (total food calorie value per day)
- Car ownership (number of cars owned per 1000 people)
- Access to safe water (measured as a percentage of the total population)
- Infant mortality (number of babies that die before one year of age, measured per 1000 live births each year)

Human Development Index

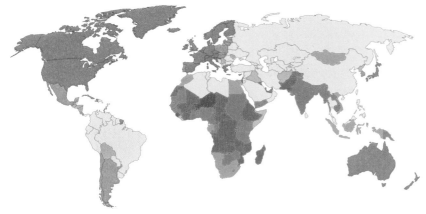

High human development	0.9–1.00
	0.8–0.89
Medium human development	0.7–0.79
	0.6–0.69
	0.5–0.59
Low human development	0.4–0.49
	0.3–0.39
	0.2–0.29
Not applicable	

Key Point

There are many ways to measure development, apart from wealth.

Classifying Countries by Economic Development and Quality of Life

- Countries can be classified according to their level of development as:
  - **LIC** (lower income country) – classified by the World Bank (in 2022) as less than $1085 GNI (gross national income) per capita, e.g. Ethiopia in Africa

- **HIC** (higher income country) – classified by the World Bank (in 2022) as more than $13 205 GNI per capita, e.g. the UK and USA
- **NEE** (newly emerging economy) – a country that has begun to experience high rates of economic development, usually with rapid industrialisation. They differ from LICs as they no longer primarily rely on agriculture. The leading NEEs are known as BRICS (Brazil, Russia, India, China and South Africa).

Limitations of Economic and Social Measures

- The use of data can give a false picture of development as it gives an average for a whole country. In some countries, such as Brazil and China, there can be great variations in levels of development between regions.
- The use of a single measure of development can be misleading, e.g. low birth rates generally suggest good social development, but birth rates can be reduced by government policy.
- Data can be out-of-date.
- Data can be unreliable due to misreporting or not available due to conflict, disasters or difficulties accessing very remote areas.
- Corrupt governments may distort the available data.

Development and the Demographic Transition Model

- The **Demographic Transition Model** (DTM) shows population changes over time. As a country becomes more developed, its population characteristics change.
- The difference between birth rates and death rates represents the natural population increase. If birth rates are higher than death rates, the total population will increase.
- The DTM is divided into five stages. Most HICs are in stage 4 or stage 5, while most LICs are in stage 2 or stage 3.
- Global population is rising rapidly because birth rates are still very high in LICs, while death rates have fallen across the world due to global efforts to tackle malnutrition and diseases such as smallpox.

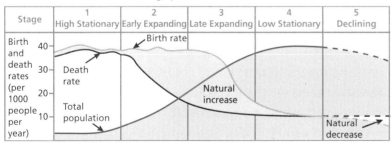

Demographic Transition Model

Key Point
Countries develop at different rates and there is a wide range of reasons for this.

Key Point
Countries with different levels of development and quality of life have different population characteristics.

Population Characteristics of HICs
- Growth rate stable
- DTM stage 4
- Low birth rates, low death rates
- Ageing population
- Some countries, like Germany, have declining populations (DTM stage 5)

Population Characteristics of LICs
- Population rising rapidly
- DTM stage 2 or 3
- High birth rates, falling death rates
- Youthful population

Key Words
development
quality of life
Human Development Index
LIC
HIC
NEE
Demographic Transition Model

Quick Test
1. Identify three development indicators (other than wealth).
2. Using the map of Human Development Index on page 76, identify two HICs, two LICs and two NEEs.
3. Who are the BRICS?

The Development Gap

Quick Recall Quiz

You must be able to:

- Identify reasons why development is uneven
- Explain the consequences of uneven world development
- Examine various strategies to reduce the development gap.

Causes of Uneven Development

Physical and Social

- Extreme climates can affect access to fresh water and the ability to grow food.
- Some countries, such as the Philippines, are affected by high magnitude and/ or frequent natural hazards (tropical storms and earthquakes) that hinder development.
- Some countries have limited natural resources, such as fossil fuels and minerals.
- Steep, mountainous terrain is harder to build on and limits farming.
- Some countries are **landlocked**, making international trade difficult. Of the 15 lowest-ranking HDI countries, eight are landlocked (e.g. Chad in Africa).
- In many of the LICs, populations are growing rapidly and putting pressure on resources. This affects access to healthcare, education, water and electricity – all of which are needed for development.

Economic and Historical

- Many LICs rely on primary resources, food stuffs, minerals and fossil fuels for export. Prices can fluctuate for these products and harvests can fail owing to natural disasters, outbreaks of disease and other sources of a mineral being found elsewhere, leaving countries with little income.
- **Transnational corporations (TNCs)** can be accused of exploiting LICs, paying low prices for raw materials and food. TNCs and some countries invest in mining activity by providing the infrastructure required, railways and ports (e.g. China's investment in lithium mining in DR Congo and Zimbabwe). Local people get low paid manual jobs while the management jobs often go to the foreign nationals. This is a form of **neo-colonialism**.
- In the 1700s and 1800s, **colonialism** saw a number of European nations exploit the resources and labour of many of today's LICs (e.g. Nigeria was a colony of the UK). This meant that countries could not take advantage of their own resources to generate their own development and some countries had thriving industries wrecked.
- Colonialism ended during the mid-20th century, meaning that many countries in Asia and Africa became independent but still relied on the old colonial power for help and assistance. Political tensions remain in many of these countries, which can result in war and conflict, and can limit development.

Key Point

There is a range of physical, social economic and historical reasons for uneven development.

Consequences of Uneven Development

Disparities in Wealth and Health

- The most developed countries have the greatest wealth and a good balance of trade, whereas the quality of life is worsening in some countries.
- Within countries there can be great disparities in wealth, e.g. in Brazil and South Africa.

- LICs have become dependent on HICs for aid, often resulting in debt.
- LICs have a shortage of safe, clean water and are unable to invest in good quality healthcare, with infectious diseases such as malaria, tuberculosis and diarrhoea the main cause of deaths – particularly among children.

International Migration

- People seek to move in order to improve their quality of life.
- Some migrants move voluntarily looking for work opportunities, higher wages, better education and healthcare.
- Some people are forced out of countries owing to conflict, persecution and poverty.

Reducing the Development Gap

- 'Development gap' refers to the difference in the standard of living and well-being between the richest and the poorest countries of the world. The gap can be reduced in these ways:

Aid can fund pumps to provide clean water

International aid	Can be **short-term aid** (which solves an immediate problem like lack of homes after an earthquake) or **long-term aid** (which tries to solve a problem so it never happens again, e.g. building a dam to provide a clean water supply). Aid can be provided by a country, a group of countries or charity groups.
Intermediate technology	Uses simple, appropriate technology that the local community can set up, work with and maintain to improve development (e.g. pumps to provide clean water).
Fair trade	Where producers receive a guaranteed fair price for the things they make and grow. A fair price is one that covers costs of production and enables people to have a reasonable standard of living. It also puts money into community projects like water and schools.
Debt relief	Between 1960 and 1980, many LICs were loaned money from HICs, banks and international organisations. This money was loaned out with high rates of interest and in some cases mis-spent or siphoned off by corrupt governments. Many LICs have been left with debts which they cannot pay back. Cancelling these debts or lowering interest rates can help LICs to develop.
Micro-finance loans	Small amounts of money lent to individuals in LICs, such as by the Grameen Bank, to help them start their own businesses. This enables people to work their way out of poverty and helps development at a local level.
Investment	Many TNCs have invested into LICs and NEEs (e.g. Apple manufactures iPhones and iPads in China). This creates jobs and puts money into the local economy.
Industrial development and tourism	Investing in manufacturing and tourism creates greater wealth than primary industries. Governments can tax this wealth, which can then be invested into public services.

Key Point

Various strategies exist for reducing the global development gap.

Kenya: Using Tourism to Reduce the Development Gap

- Kenya attracts 700 000 visitors each year owing to: wildlife safaris (to see elephants, rhinos, buffalos, lions and leopards); the tribal culture; a warm climate with sunshine all year; varied scenery, including savannah, grassland, mountains, forests, beaches and coral reefs; 23 national parks, e.g. Tsavo and Masai Mara.
- Tourism accounts for 15% of Kenya's GDP.
- TNCs, such as Hilton Hotels, have invested in Kenya.
- Tourism creates jobs directly in hotels and indirectly in supporting industries like taxi companies. There are 250 000 tourism-related jobs.
- Tourists spend money with local businesses, which boosts the economy.
- Infrastructure is improved to cope with the influx of visitors.

Safaris are big business in Kenya

Key Words

landlocked
transnational corporation (TNC)
colonialism
short-term aid
long-term aid
intermediate technology
fair trade

Quick Test

1. Why does international migration occur?
2. How does long-term aid differ from short-term aid?
3. What is the main principle of fair trade?

Changing Economic World Case Study – Vietnam

You must be able to:

- Describe the changing industrial structure of a particular location and its context
- Examine the role of transnational corporations, political and trading relationships, and international aid
- Explain the impacts of economic development on people and the environment.

Location, Importance and Wider Context

- Vietnam is a country located in south-east Asia (bordering China, Laos and Cambodia). Its capital city is Hanoi. It is a communist country.
- Vietnam is a former French colony. It became independent in 1945.
- Vietnam has a population of 97.3 million (2020). It is the world's 15th most populated country. The population is growing steadily at a rate of 0.93%. The birth rate is 16 per 1000 and the death rate is 6 per 1000. The country has a youthful age structure.
- 35% of the population live in urban areas, although this is changing with rapid rural-to-urban migration.
- Vietnam has a tropical climate in the south, with hot temperatures and high rainfall all year round. In the north, the climate is monsoonal with a warm rainy season May to September and a dry season October to April.
- Vietnam has a range of natural resources, such as coal, manganese, timber, and offshore oil and gas deposits.
- Vietnam has changed from being a low-income country (LIC) to a newly-emerging economy (NEE). Since 2000, it has had one of the fastest rates of economic growth in the world.

Changing Industrial Structure

- Despite rapid economic growth in Vietnam, 37% of people still work in primary sector jobs such as farming, mining and forestry. The main crops are rice, coffee, tea, soybeans, peppercorns, cashew nuts, fruits and rubber.
- The secondary sector has grown and now employs 24% of people and generates 33% of the country's GDP. This sector is growing rapidly because Vietnam is an important offshore location for **transnational corporations (TNCs)**.
- The service sector is growing rapidly and now employs 35% of people and generates 51% of the country's GDP.
- The tourist industry is growing rapidly in Vietnam with over 4 million international visitors a year.

Role of Transnational Corporations

- Owing to cheap labour, a growing home market and fewer industrial laws and restrictions, TNCs are investing in Vietnam.
- Nike, the sportswear and goods manufacturer, has 34 plants in Vietnam (75% of its workforce is based in south-east Asia). The majority of workers are women under 25.

Advantages of TNC Investment

- Creation of jobs
- The wages that workers earn are spent in the local economy, creating a positive multiplier effect
- Investment from foreign companies brings new technology and ideas
- The government earns tax revenues, which can be invested into infrastructure

Disadvantages of TNC Investment

- Political influence held over the Vietnamese Government as investment can easily be moved away from Vietnam to a lower cost location
- Many of the jobs created are low paid, especially when compared to jobs in HICs
- Some profits from TNCs return back to the home country – **economic leakage**
- New industries can have an environmental impact, including pollution from factories

Political and Trading Relationships

- Vietnam is becoming economically more important. It has a high annual GDP growth rate of 6.3%, representing its strong manufacturing exports, and rising demand for goods and services within the country.
- By 2050 it is predicted that Vietnam will overtake countries like Portugal, Norway and Singapore in terms of total GDP.
- Vietnam has a trade surplus, meaning that it exports more than it imports.
- Exports include crude oil, clothing, seafood, rice, tea, coffee and electronics. Export destinations include USA, Japan, China, South Korea and Australia.
- Imports include machinery, steel products, raw materials for the clothing and footwear industries, electronics, plastics and cars. The main import partners are China, South Korea, Japan, USA and Thailand.
- Vietnam joined the World Trade Organisation in 2007 and the Association of Southeast Asian Nations (ASEAN) economic community in 2015, showing its growing influence at a regional and global scale.

> **Key Point**
>
> Countries like Vietnam are newly-emerging economies and see manufacturing industry and the export trade as ways to help reduce the development gap.

International Aid

- Vietnam receives multilateral aid from the World Bank. This is to help sustainable development and to promote good governance.
- Bilateral aid is given to Vietnam from countries such as Australia and the USA. In 2017–2018 Australia was to provide $484.2 million for projects such as improving transport infrastructure, training workers and promoting women's economic empowerment.
- Voluntary aid is being given to Vietnam by charities, e.g. Action Aid has been working on women's rights, tackling hunger and providing education.

Effects on Environment and Quality of Life

- Industrialisation and urbanisation (e.g. in Hanoi and Ho Chi Minh City) are causing rising levels of air and water pollution, and waste.
- Logging and farming are causing deforestation with resulting loss of animal habitats, increases in flood risk and soil erosion.
- Overfishing is threatening marine life populations.
- Energy consumption has risen with economic development. 58% of electricity is generated by fossil fuels, which release greenhouse gases.
- Growth in the economy has seen people's incomes rise significantly.
- Poverty levels decreased from 58% in 1993 to 3.2% in 2016.
- Having more money means people can improve their quality of life, e.g. access to clean water, healthcare and better housing.
- There are significant inequalities in quality of life, with people in rural areas and from ethnic minorities more likely to live in poverty.
- It has been reported that some people (e.g. those employed by TNCs) have to work long hours in difficult conditions.

Forest cleared for agriculture in the Vietnamese highlands

Quick Test

1. Identify three reasons why a TNC might want to locate in a country like Vietnam.
2. What is the difference between multilateral and bilateral aid?
3. List three negative environmental impacts resulting from economic development in Vietnam.

> **Key Words**
>
> economic leakage

UK Economic Change

Quick Recall Quiz

You must be able to:

- Understand the causes of economic change in the UK
- Explain what is meant by the north–south divide and be aware of strategies that have been used to resolve regional differences
- Discuss the social and economic changes in rural landscapes; in an area with population growth and in an area with population decline.

The Changing UK Economy

- The economic structure of the UK has changed over time.
- Since the 1980s, there has been a decrease in primary and secondary employment and this is known as **deindustrialisation**.
- Primary employment, such as farming and mining, has declined due to increased mechanisation and automation, meaning that fewer workers are required, and the depletion of locally available raw materials.
- Secondary employment, such as the manufacturing of clothes, has declined owing to competition from overseas where labour costs are cheaper and production methods are more advanced.
- Government policy, such as investment in infrastructure like roads and industrial estates, financial incentives and faster planning permission, has attracted some industry to the UK (e.g. Nissan, the Japanese car manufacturer, opened a car plant in Sunderland).
- Since 2001, there has been an increase in tertiary employment, e.g. tourism, financial services and healthcare.
- Since 2001, there has also been an increase in quaternary employment – 'knowledge-based' industries such as software companies, research and development, and biotechnology in science and business parks on the edges of cities like Bristol and Cambridge.
- In recent years there has been an increase in flexible part-time working, job sharing and home teleworking using the Internet (which accelerated during the Covid-19 pandemic).
- The economic structure varies around the UK, with different areas specialising in different industries (e.g. London is a financial centre).

Positive Impacts of Globalisation on the UK Economy

- Cheaper goods and services as they are produced in places where labour costs are lower
- Investment from foreign companies bringing new technology and ideas
- Migrants fill jobs where there is a shortage of workers, e.g. agriculture and construction

Negative Impacts of Globalisation on the UK Economy

- Business closures and loss of jobs as companies close down and move overseas, where labour costs are cheaper
- Loss of skilled British workers who migrate overseas for higher wages
- Greater pay inequality between highly skilled and unskilled workers

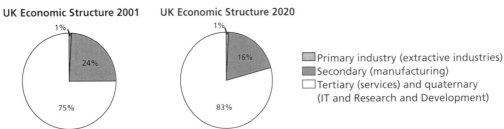

UK Economic Structure 2001

1%
24%
75%

UK Economic Structure 2020

1%
16%
83%

- Primary industry (extractive industries)
- Secondary (manufacturing)
- Tertiary (services) and quaternary (IT and Research and Development)

Key Point

There has been a shift away from traditional forms of employment in recent years.

The North–South Divide

- The UK demonstrates many regional patterns.
- The North has depended on heavy industries like steel-making and ship-building, which have suffered a decline in recent years, while London and the South-East have developed rapidly owing to a fast-growing service sector, e.g. finance, high technology, research and media.

- Wages are higher in the South. Unemployment is higher in the North.
- Average house prices are higher in the South owing to higher demand.
- Population is growing more rapidly in the South owing to more work opportunities and higher wages.
- There are exceptions to the north–south divide and there are large inequalities within particular areas, such as in London.

Strategies to Resolve Regional Differences

- The Government has created the 'Great Northern Powerhouse' initiative.
- The Government has agreed devolution measures to give extra power and money to directly-elected mayors to ensure decisions affecting the North are made by the North.
- Improvements to transport infrastructure linking cities in the North, making it a more attractive place to live and work. Manchester will be connected to HS2, a high-speed rail link between the UK's major cities that is due to open in 2026.
- The Government has established **enterprise zones**, providing tax breaks and government support. They have attracted new and expanding businesses, creating jobs and supporting economic growth.

Social and Economic Change in Rural Areas

An Area with Population Growth

- **Counterurbanisation** is where people move from urban areas to surrounding rural areas for a better quality of life. They then typically commute to work in a nearby urban area.
- Bramhall is a village 10 miles south of Manchester. It is a popular place in which to live, having a range of housing, services and good transport connections. Bramhall has its own railway station, is near to motorways such as the M56 and M60, and is close to Manchester Airport.
- In 1911, the population of Bramhall was approximately 2500; by 2011 it had increased to around 17 500.

An Area with Population Decline

- The Outer Hebrides is a set of islands off north-west Scotland.
- It is an extremely rural area with a population density of 9 people per km^2.
- In 1911, the population of this area was approximately 45 000; by 2011 it had gradually fallen to about 30 000.
- The economy of the area is largely based on farming and fishing, meaning there are few job opportunities. There is limited transport availability and housing opportunities.
- Young people have moved away to find job opportunities, leaving behind an ageing population.
- Owing to population decline, there is less demand for shops and services, resulting in business closures. Fewer people also mean that the local council has less money from taxes to invest in public services like healthcare and schools. These services get reduced and close. The area is not attractive to inward investment, so buildings are left derelict. As the quality of life deteriorates, more people move away; this is known as a **negative multiplier effect**.

Positive Impacts of Population Growth on Bramhall

- Population growth has supported local services and led to new shops and services opening
- New local businesses have started up
- Old housing has been modernised
- There is a good community spirit with lots of community events throughout the year

Negative Impacts of Population Growth on Bramhall

- House prices are pushed up, forcing some people out of the area
- Services become orientated towards wealthy people rather than meeting the needs of the local community
- Local schools are oversubscribed
- At commuting times there is heavy congestion on local roads, causing noise and air pollution

Key Point

Population change has caused major economic and social changes in rural areas.

Key Words

deindustrialisation
enterprise zone
counterurbanisation
negative multiplier effect

Quick Test

1. Identify three reasons for deindustrialisation in the UK.
2. What is an 'enterprise zone'?
3. List two economic and two social impacts of population growth in a rural landscape.

UK Economic Development

You must be able to:

- Describe different types of links between the UK and the wider world
- Outline improvements and new developments in transport
- Explain how industry impacts on the environment and suggest ways in which industrial development can be made more sustainable.

The Place of the UK in the Wider World

- **Globalisation** is the process by which the world is becoming increasingly interconnected in terms of industry, markets, migration and cultures. The UK has strong links with the wider world.
- The UK trades with countries around the world, particularly those in Europe, which are geographically close and economically wealthy. Since leaving the European Union in 2020, the UK has been negotiating new trade deals with countries around the world.
- Investment by individuals and firms from abroad (known as foreign direct investment) has made a major contribution to the UK economy. For example, Honda, the Japanese car manufacturer, has a plant in Swindon.
- The UK has strong creative industries – music, television, film, etc. These products are exported worldwide.
- The Channel Tunnel links the UK to France and, therefore, mainland Europe. There are a number of international airports in the UK. Transport links enable people and goods to move between countries.
- People and businesses in the UK have very good access to telecommunication and Internet networks, enabling communication, marketing and financial transactions to take place at a global scale.
- The UK works closely with Commonwealth countries, an association of 53 independent states, many of which were part of the former British Empire. They share values of democracy, human rights and the rule of law. The UK is linked to many of these countries through trade, culture and migration.
- The UK is part of the G7 (the seven most powerful industrialised countries) with the USA, Canada, France, Germany, Italy and Japan. These countries are very important in global decision-making.
- The UK regularly hosts internationally important sports, music and cultural events.

> **Key Point**
>
> The UK is a major player in the global economy.

UK Transport Developments

- **Infrastructure** improvements in the UK include road building and improvement programmes. In some places, motorways are being widened with an extra lane. Many motorways (e.g. the M6) are being made 'smart', whereby speed limits are varied to try to keep traffic moving and to reduce congestion.
- There has been government investment in railways. A high-speed rail link runs from London to the Channel Tunnel. Crossrail 1 and 2 are being constructed to improve transport in the London area.
- The HS2 link should reduce pressure on roads and rail networks and will significantly cut journey times between major cities.

A high-speed train operating between London and the Channel Tunnel

- UK airports are operating at near-capacity. There are proposals to expand London Heathrow, the UK's largest airport, by building a third runway. There has also been talk of building a second runway at Manchester Airport, helping to boost the economy in northern England.
- A new port on the Thames estuary, called London Gateway, opened in 2013. It can accommodate large container ships and enables goods to be brought close to London, the largest consumer market in the UK.

Impacts of Industry on the Environment

- All types of industry can have negative effects on the environment.
- Primary industries have a huge impact on the environment. Farmers use herbicides, pesticides and fertilisers, which can run off into water sources and impact supplies and ecosystems. Mining and quarrying create lots of dust and noise pollution and can leave permanent scars on the landscape.
- Manufacturing industries construct large factory buildings that can destroy **greenfield sites** and intrude on the landscape.
- Tertiary and quaternary industries create lots of waste, which can pollute the landscape and the air through waste-disposal processes.
- Most industries use lots of electricity, which is often created by burning fossil fuels. This creates greenhouse gas emissions, contributing towards the enhanced greenhouse effect and, therefore, climate change.
- Transport of raw materials and finished products for different industries (usually by road) increases noise and air pollution.
- Traditional UK industries often caused a lot of pollution. Modern industry is being made more environmentally sustainable through:
    - use of modern technology (like desulphurisation) to reduce emissions from factories and power stations
    - use of renewable sources of energy
    - investment into energy efficiency
    - stricter environmental targets for water quality, pollution and landscape damage
    - heavy fines for industrial pollution
    - choosing **brownfield sites** for development.
- Adnams is a brewery producing beers and spirits based in Southwold in Suffolk. Since 2017, the business has switched entirely to renewable sources of electricity – wind, solar and hydro. This has massively reduced the company's carbon emissions and, therefore, its contribution to the enhanced greenhouse effect and climate change.
- Adnams is trying to reduce its water consumption, which is very important given that Southwold is located in the UK's driest region. The business is working hard to reduce packaging. It has reduced the amount of glass used in beer bottles, and it charges customers for carrier bags and donates the proceeds to the Suffolk Wildlife Trust. The company tries to ensure that buildings blend in with the landscape as best as possible and tries to ensure that biodiversity is preserved.

Road transport is a major source of pollution

 Key Point

Industrial development can lead to a wide range of negative impacts on the environment.

Adnams Brewery Distribution Centre with a green roof, solar panels and lime/hemp walls

 Key Words

globalisation
infrastructure
greenfield site
brownfield site

Quick Test

1. What is meant by the term 'globalisation'?
2. List two benefits of the Government's investment into high-speed railway networks.

Review Questions

Coasts 1: Processes

1 What is meant by a 'constructive wave'? [2]

2 How might a beach change if there are a lot of surging waves? [2]

3 Describe hydraulic action and the results. [3]

Total Marks _____ / 7

Coasts 2: Landforms

1 How are wave cut platforms formed? [3]

2 Explain why headlands are eroded but bays become places of deposition. [4]

Total Marks _____ / 7

Coasts 3: Management

1 Why is a beach a good form of coastal protection? [3]

2 Why are sand dunes easily damaged by humans? [3]

3 Using the diagrams below, describe some of the changes resulting from building at the coast.

Before building

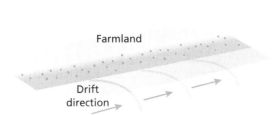

Farmland

Drift direction

After building

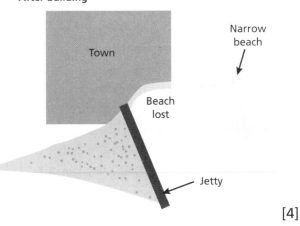

Town

Narrow beach

Beach lost

Jetty

[4]

Total Marks _____ / 10

Rivers 1: Processes

1 Explain why a particle of sand is easier to entrain than a particle of silt or clay. [2]

2 Describe what could happen to the river bank in the diagram as a result of abrasion. [3]

3 Explain two reasons for deposition taking place in rivers. [4]

> **Total Marks** _____ / 9

Rivers 2: Landforms

1 Why do spurs often 'interlock' in upland and small rivers? [3]

2 On the diagram below, indicate the location where a waterfall is likely to form. [1]

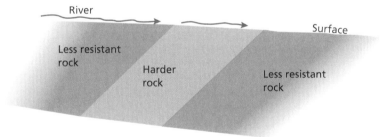

> **Total Marks** _____ / 4

Rivers 3: Flooding and Management

1 What is a storm hydrograph? [2]

2 What effect will a very high water table have on the ability of a channel to hold water? [3]

3 Referring to a specific place or flood event, explain **one** human influence on flooding. [3]

4 Describe two **natural** reasons why a specific place flooded or might flood. [4]

> **Total Marks** _____ / 12

Review Questions

Glaciation 1: Processes

1 Where is supra-glacial moraine? [1]

2 The diagram below shows a glacier transporting material.

What types of moraine are represented by A, B and C?

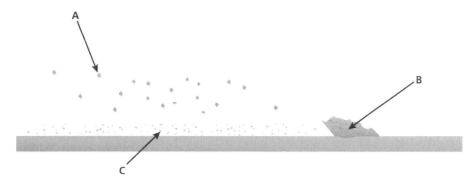

[3]

3 Why do glaciers deposit material? [4]

> Total Marks .... / 8

Glaciation 2: Landscape

1 Where would you expect to see a ribbon lake? [1]

2 Why would a valley with lots of tributaries become a very deep glaciated valley? [2]

3 Describe how abrasion shapes a corrie. [3]

> Total Marks .... / 6

Glaciation 3: Land Use and Issues

1 Describe the climate of a named upland glaciated area in the UK. [4]

2 How can farming in glaciated uplands damage the environment? [4]

> Total Marks .... / 8

Urbanisation

1. What percentage of the world's population lived, or is predicted to live, in urban areas in the following years? Choose from the following possible answers: **10% 25% 42% 56% 68%**

   a) 1900   [1]   b) 2020   [1]   c) 2050   [1]

2. What does the term 'urban sprawl' mean, and why might it be considered a problem?   [3]

3. a) List some of the different push factors that lead to rural-to-urban migration.   [3]

   b) List some of the different pull factors that lead to rural-to-urban migration.   [3]

4. Explain the differences in urban growth between the richer and poorer parts of the world.   [4]

> **Total Marks** _____ / 16

Urban Issues and Challenges 1

1. Explain why so many rural people are attracted to cities like Rio de Janeiro in Brazil.   [4]

2. Rio de Janeiro has a large number of shanty towns. What problems are faced by the people who live in them?   [4]

3. Tourism is a major industry in Rio de Janeiro. Describe some of the attractions the city has for visitors.   [4]

4. Describe in detail two social challenges of life in Rio de Janeiro.   [4]

> **Total Marks** _____ / 16

Urban Issues and Challenges 2

1. Give two ways in which sites along the River Thames have been re-imaged.   [2]

2. Identify four social improvements that formed part of the legacy of the London Olympics.   [4]

3. Explain two ways in which transport systems can be improved in London to help reduce carbon emissions.   [2]

4. Describe two new projects that aim to improve transport networks in London.   [4]

> **Total Marks** _____ / 12

Urban Issues and Challenges 3

1. What percentage of the UK population live in urban areas? [1]

2. Cities can sometimes be described as 'systems'. Name two inputs into a city system. [2]

3. Describe in detail three different strategies that can help to reduce traffic congestion. [6]

4. What changes can be made to modern cities to improve resource management? [6]

> **Total Marks** _____ / 15

Measuring Development and Quality of Life

1. Name two social indicators that could be used to measure development. [2]

2. Name two economic indicators that could be used to measure development. [2]

3. Identify four population characteristics of a HIC. [4]

> **Total Marks** _____ / 8

The Development Gap

1. Describe two physical causes of global inequalities. [4]

2. How can 'fair trade' help to improve life for farmers in LICs? [4]

3. What attractions can Kenya offer to bring in tourists? [4]

4. How can tourism help to reduce the development gap in Kenya? [4]

> **Total Marks** _____ / 16

Changing Economic World Case Study – Vietnam

1 What factors attract many TNCs to locate in NEEs? [3]

2 Explain what is meant by 'economic leakage'. [1]

3 Describe the industrial structure of a NEE you have studied. [3]

> **Total Marks** _____ / 7

UK Economic Change

1 Name two cities where science parks have developed. [2]

2 Complete the following table of employment data in the UK:

2001	2020 – higher, same or lower?
Primary = 1%	
Secondary = 24%	
Tertiary and Quaternary = 75%	

[3]

3 Outline strategies that have been used to reduce the north–south divide in the UK. [3]

> **Total Marks** _____ / 8

UK Economic Development

1 List two of the UK's main trading partners. [2]

2 Explain how industrial development can be made more sustainable. [3]

3 Distinguish between a greenfield site and a brownfield site. [2]

> **Total Marks** _____ / 7

Overview of Resources – UK

You must be able to:

* Understand how importing food products from abroad can increase carbon footprints and food miles
* Explain how demand for water is increasing in the UK
* Appreciate how the sources of energy used to power the UK are changing.

Food in the UK

* Only around 25% of all the fruit and vegetables consumed in the UK are grown there.
* British supermarkets sell fruit and vegetables like green beans, mangetout, peas and strawberries that are imported from abroad, especially from countries that have climates allowing these products to be grown all year.
* Most **organic** produce sold in UK supermarkets is grown within the UK but is only sold when 'in season'.
* The import of fruit and vegetables from countries like Kenya and Peru leads to increased **food miles** and **carbon footprints** but does help the economies of those countries.
* There is an increasingly popular trend to only eat locally produced meat, fruit and vegetables only when in season, thereby cutting down on food miles.
* In the UK, agribusiness has become more important in the last 40 years as farm and field sizes have increased along with the use of machinery and agrichemicals, while the number of farm workers has decreased.

Key Point

The UK is reliant on imported food from around the world; a process that increases its carbon footprint.

Water Use in the UK

* With increasing affluence, water demand is growing as new houses are built with more than one bathroom and consumers demand labour-saving devices such as dishwashers.
* Industry and agriculture are also large users of water, e.g. 15 000 litres of water are needed to produce 1 kg of beef.
* Maintaining the quality of drinking water and the management of water pollution are costly, requiring great investment in technology and with costs that are largely met by the consumer.
* The areas of highest demand in the UK – London and the South East – are the areas where there is least available water.
* The areas with least demand for water – the north and the west of the UK – are also the wettest and have surplus water.
* To maintain water supplies, water is transferred from wetter regions to those drier areas with greatest demand. For example, water is sent from Kielder Water in Northumberland to cities like Leeds using a system of pipelines, aqueducts and rivers.

Key Point

Increasing water consumption is leading to water supply problems in some parts of the UK.

Energy Use in the UK

- In the 1960s, most of the UK's electricity was generated using coal, oil and a small amount of nuclear (gas was also made from coal).
- Gas and **renewables** now generate most of the UK's electricity, with nuclear and bioenergy (generated from organic matter such as plants, wood and food waste) also producing a significant share.
- As domestic supplies of coal, gas and oil have reduced, the UK has become more dependent on foreign sources of fossil fuels, the cost of which has been increasing with higher global demand.
- Some see nuclear power as a way of reducing the UK's dependence on fossil fuels as it is a low carbon alternative, but it has many disadvantages (such as the toxic waste it creates).
- The use of renewables is growing in significance in the UK as the number of wind and solar power installations increases.
- The development of renewable energy is expensive and the cost to the consumer can be higher than electricity from fossil fuels.
- New methods of exploiting energy sources include shale gas that is obtained through **fracking**, a hydraulic fracturing process that has many environmental issues associated with it.

UK energy production, 2021 (approx.)	
Gas	37%
Wind	26%
Nuclear	14%
Bioenergy	12%
Solar	4.5%
Coal	2%
Hydro	2%
Other	2.5%

Key Point

Alternative sources of energy will become increasingly important in the future.

Fracking

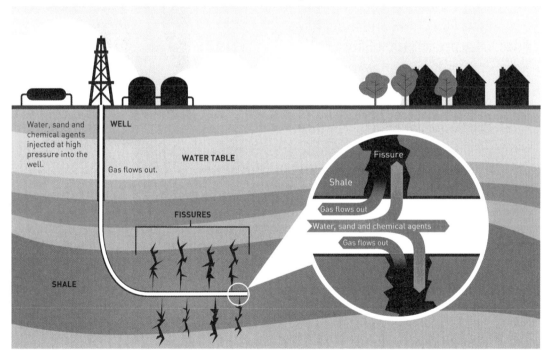

Quick Test

1. Why does the UK need to import food from abroad?
2. Why is water demand increasing in the UK?
3. Why do London and the South East have problems in supplying enough water?
4. What challenges does the UK face to provide enough energy in the future?

Key Words

organic
food miles
carbon footprint
renewables
fracking

Quick Recall Quiz

Food 1

You must be able to:

- Understand that the global demand for food is rising but supply can be insecure, which may lead to conflict
- Explore the reasons for increasing food consumption and factors affecting food supply.

Food Security

- In many developed countries, **food security** is the norm and the population are well-fed and nourished.
- More than 800 million people, mainly in developing countries throughout the world, live with hunger or **food insecurity**.
- Food production is influenced by two factors – the physical environment and human capacity – and is not evenly distributed throughout the world.
- Climate, water availability and soil type are the main components of the physical environment, while human capacity refers to population size, the farming skills and the financial investment a country is able to put into its agriculture.
- Countries with vast populations like China and India have agricultural output values in excess of $100 billion.
- HICs such as the USA have very large output values due to the intensification of farming methods and large investments of capital.
- In contrast, LICs, like those found in some parts of Africa, produce less food.
- Calorie intake also varies between rich and poor countries.
- People in the USA and EU consume the most calories per capita on average (3000 per day).
- This is much higher than the recommended daily calorie intake of 2500 for men and 2000 for women, leading to obesity in these countries.
- Some African countries are experiencing below average per capita calorie intake.

Packages of ripe red vine tomatoes on the production line in a food processing plant

Increasing Food Consumption

- Many LICs have rising populations and, as the world's population increases by around 80 million people each year, more land is put under cultivation to feed them.
- As countries develop economically, their population demands a more western-style diet with more meat and dairy products.
- This leads to grain that was once used to feed the people being fed to animals, resulting in shortages for the poor and diet-related illness for the rich.

> ### Key Point
>
> HICs have high levels of food security while LICs tend to have low levels.

Factors Contributing to Food Insecurity

- Extreme climatic conditions (such as drought or floods) can lead to low agricultural **yields** and environmental degradation (such as desertification), in turn leading to **famine**.
- Famine is more complex than just a shortage of food. The UN definition of famine is complex and has three conditions:
  - More than 20% of the population must have fewer than 2100 calories of food available per day.
  - More than 30% of children under the age of 5 must be acutely malnourished.
  - Two deaths or more per day in every 10 000 people or four deaths per day in every 10 000 children must be being caused by lack of food.
- Famine causes rising food prices. Other factors can also cause rises to food prices, such as natural disasters like drought, leading to under-nutrition and starvation among the poorest in those countries.
- **Climate change** has also had an impact in some of the countries of Africa, as soil erosion and desertification have led to a lack of land to grow food on.
- The countries that face the biggest food shortages, and therefore the worst levels of food insecurity, are often those that face the biggest threats from agricultural pests and diseases.
- Those countries also face the biggest risk from water shortages caused by unpredictable rainfall.
- The Sahel region of Africa has been suffering from drought on a regular basis since the early 1980s. The 1983–85 famine in northern Ethiopia was the worst to hit the country in a century and led to 400 000 deaths.
- The effects of the famine were made worse by a background of social unrest and civil war, which had raged in Ethiopia and Eritrea for two decades.
- Famine caused by civil war in Yemen has led to 9 million people (the population of London) being affected by food insecurity.

Poor physical conditions can contribute to food insecurity

 Key Point

Producing food sustainably is a challenge for most nations.

Quick Test

1. What is food security?
2. Suggest physical reasons for food insecurity.
3. Suggest human reasons for food insecurity.

 Key Words

food security
food insecurity
yield
famine

Food 2

You must be able to:

- Describe the range of strategies that increase food supply
- Explain the ways in which food production can be made more sustainable.

Increasing Food Supply

- **Irrigation** is the application of water to increase the yields of crops, especially in those areas where water is in short supply.
- It is also used in HICs where the plentiful supply of water can be seasonal.
- The Gezira Scheme in Sudan is an example of a large-scale irrigation project, where the floodwaters of the Nile are used to grow crops such as cotton and wheat. However, large amounts of water are taken out for this process, which affects other irrigation schemes downstream.
- Hydroponics is a method of growing plants using nutrient-rich water, often using inputs of ultra-violet light; for example, growing high value crops like tomatoes in greenhouses.
- Aeroponics is a similar way of growing plants in an air or mist environment, without the use of soil.
- The Blue Revolution refers to the growth of aquaculture as a very highly productive way of producing food, such as aquatic animals and plants in both saltwater and freshwater.
- Fish farming, such as shrimp or salmon farming, and the gathering of seaweed are all methods of aquaculture.
- The Green Revolution is an initiative that has increased agricultural production, particularly in LICs, since the late 1960s.
- The Green Revolution saved around a billion people from starvation and involved the development of high-yielding varieties of cereals (rice and wheat) and new farming techniques, including the use of irrigation, synthetic fertilisers and pesticides.
- Biotechnology uses biological processes to develop new products that can help feed the hungry by producing higher crop yields.
- Biotechnology can lower the input of agricultural chemicals into crops, resulting in fewer vitamin and nutrient deficiencies and reduced allergens and toxins.
- Appropriate technology is small-scale technology that is simple enough that people can manage it directly and on a local level.
- It makes use of skills and technology that are available in a local community to supply basic human needs, such as gas and electricity, water, food, and disposal of waste. Examples include the Jiko Stove, which is made from locally sourced materials.

Irrigation

Key Point

Technology can be used to increase food supply and reduce food insecurity.

Jiko Stove

Making Agriculture More Sustainable

- Organic farming is a sustainable method of food production that does not use synthetic pesticides or fertilisers. Organic products, such as plant and animal waste, are used as fertiliser and biological solutions have been developed for pest and disease control.
- Urban farming is a sustainable practice where growers cultivate and process produce, often for their own use, and can include a variety of activities such as vegetable and fruit growing or beekeeping.
- In LICs, where poverty is a problem, urban farming is often practised by urban dwellers to produce enough food for their families.
- The Organopónicos in Havana, Cuba, is a good example of a local scheme to increase food security for the urban poor.
- The scheme started in the early 1990s when the Soviet Union collapsed and could no longer provide food to Cuba. The Organopónicos consist of low-level concrete walls filled with organic matter and soil, with lines of drip irrigation laid on the surface.
- Around 35 000 hectares of Organopónicos are found in Havana, growing various fruit and vegetables.
- In HICs, allotments and smallholdings are popular leisure pastimes as well as producing plentiful and cheap, nutritious food.
- **Ethical consumerism** is the purchase of sustainable goods, such as sourcing food that is locally grown and only eating fruit and vegetables that are in season.
- Other examples include changing diets to those less reliant on meat and dairy products, or using sustainable fish sources such as pollock instead of cod (ensuring that the cod will recover in overfished areas).
- Reduced food waste and losses in both homes and in shops could save UK consumers up to £2.4 billion a year.
- The average UK family throws away enough food for around six meals a week, so buying less food in the supermarket and using goods before they go 'off' would save the average family around £60 a month.
- **Permaculture** is a method of agriculture where humans work with nature to farm sustainably.
- Sustainable management of the tropical rainforest often involves permaculture to harvest products such as wild game, nuts and berries.

Key Point

Sustainable use of food resources can also lead to increased food supply and reduced food insecurity.

Throwing away food is a waste of money

Quick Test

1. How does technology increase crop yields?
2. How can changing our eating habits ensure that we can reduce our carbon footprint?
3. What is ethical consumerism?

Key Words

irrigation
ethical consumerism
permaculture

Quick Recall Quiz

Water 1

You must be able to:

- Explain why some areas have water surpluses and some have deficits
- Describe how and why water use varies in different countries
- Consider the impacts of water insecurity.

Water Supply

- Demand for water resources is rising globally but supply can be insecure, which may lead to conflict.
- The supply of water is uneven because the distribution of rainfall varies from place to place.
- Regions of the world that have water surplus have **water security**, while those with a deficit have water insecurity.
- For example, the UK, France and Germany have water security; Turkey, Israel and Jordan have water insecurity.
- Some regions have a **physical water scarcity** through natural reasons for low water supply, e.g. low rainfall in deserts.
- **Economic water scarcity** is low supply caused by human reasons, such as the lack of a water infrastructure that can be found in many LICs.
- The wettest places in the world, such as rainforests, do not support many people, while the number of people living in **arid regions** is increasing and putting a greater demand on available resources.
- As the world's population rises so does the demand for water for drinking, bathing, agriculture and industry.

Water Usage

- With economic development, the demand for water rises because domestic use increases through more labour-saving devices like washing machines and dishwashers.
- The average North American uses 570 litres of water per day, the average European uses 130 litres, and the average African around 20 litres.
- With economic development, industrial and agricultural use of water also increases due to a greater demand for goods and services, such as meat and dairy products.
- Regions that experience low rainfall will often have limited **groundwater** and an absence of surface water supplies, which in turn can lead to reduced water availability.
- Certain types of rock (such as limestones and sandstones) store water. Other types of rock (such as clays and granite) only allow for the storage of water on the surface.

> **Key Point**
>
> Water use increases with rising levels of development and population growth.

- Increasing human water usage has resulted in groundwater supplies – a non-renewable resource – being depleted, while rivers are often diverted for agricultural and industrial purposes.
- In many LICs, regions with limited water supplies lose up to 75% of water used in irrigation through leakage, evaporation or run-off.
- Fresh water is very rare – only 3% of the world's water is fresh, with two-thirds of that stored in ice caps and glaciers and unavailable for human use.
- Agriculture consumes more water than any other user and wastes much of that through inefficient management.
- It also accounts for one of the most common types of water pollution, from agricultural chemicals.
- Sources of water pollution also come from factories, vehicle emissions and human effluent.

Key Point

Agriculture is the biggest user of water.

Impacts of Water Insecurity

- Some 1.1 billion people worldwide lack access to safe drinking water, and a total of 2.7 billion find water scarce for at least one month of the year.
- Poor sanitation is a problem for 2.4 billion people as they are exposed to diseases, such as cholera, typhoid and other **waterborne diseases**.
- 1.6 million people, one-third of whom were children, died from diarrhoea in 2017.
- Many rivers, lakes and aquifers are drying up or becoming too polluted to use and more than half of the world's wetlands have disappeared.
- Climate change is altering patterns of rainfall around the world, causing shortages and droughts in some areas and floods in others.
- This has serious implications for agriculture, which could lead to food shortages and famine in areas of the world like the Sahel region of Africa.
- Over-extraction of water for energy production and industrial processes can lead to shortages and a possible source of conflict between countries.
- By 2030, more than half of the world's population may face water shortages.

Key Point

Over 1 billion people do not have access to safe drinking water.

Key Words

water security
physical water scarcity
economic water scarcity
arid region
groundwater
waterborne disease

Quick Test

1. Define 'water security'.
2. Why are water supplies in some places becoming scarce?
3. Why is access to clean water difficult in some LICs?

Water 2

You must be able to:

- Describe schemes to transfer water between regions with surpluses and those with deficits
- Explain strategies to obtain and conserve water sustainably.

Increasing Water Supply

- In countries where water is scarce, diverting supplies from one river system to another is widespread.
- The building of dams and reservoirs also increases the supply of water for multiple uses by the population.
- **Water transfer schemes** such as the South–North Water Transfer Project in China will enable 44.8 billion cubic metres of water per year to be transferred from the Yangtze River in southern China to the Yellow River Basin in arid northern China.
- The cost of the project is over $60 billion, but it is likely to cause water shortages in some parts of China and pollution in some rivers.
- As climate change makes droughts more common, a growing number of countries are turning to **desalination** (removing salt from either seawater or groundwater to make it drinkable).
- Even London now has a seawater desalination plant as the South East is a water-stressed area, especially during dry years.

> ### Key Point
>
> In countries where there is a water shortage, water is transferred from regions with surpluses to those with deficits.

Sustainable Water Conservation

- Water can be conserved through sustainable management strategies.
- **Water conservation** methods can take place in the home and garden and can include low-flush toilets, taking short showers and only using washing machines and dishwashers when full.
- In the garden, methods can include growing drought-resistant plants that need little watering, spreading mulch or composted bark on flower borders and connecting a water butt to drain pipes and using the rainwater to water plants.
- Groundwater makes up nearly 30% of all the world's freshwater and is often used in places where there is not enough water to drink.
- Groundwater is often found in porous rocks deep underground in reservoirs called **aquifers**.
- Groundwater quality is usually very good and needs less treatment than river water to make it safe to drink, as the rocks through which the groundwater flows help to remove pollutants.

A water butt

- Groundwater also responds slowly to changes in rainfall, and so it stays available during the summer and during droughts, when rivers and streams have dried up.
- Groundwater is relied on in many developing countries, especially those in Africa, because it can often be found close to villages.
- Extracting groundwater is relatively inexpensive as it is often drawn from wells – this does not require much technology and means that reservoirs are not needed to store the water before it is used.
- Aquifers are often slow to recharge (fill up) so may not always be sustainable.
- Water used in the home can be recycled by treating it in waste water (or sewage) plants.
- **Greywater** harvesting is a way of conserving water, involving the recycling of water used in baths and showers, as well as rainwater from roofs, to be used for flushing toilets and other non-drinking water purposes.
- Many modern buildings in the UK use greywater for these purposes.
- Local sustainable water schemes are found in many LICs, however many are beset with problems, such as the Agra clean water project in northern India.
- This scheme is designed to provide water to the poorest areas of Agra, where waterborne diseases are a persistent problem due to polluted sources.
- The solution has been partially solved by bottling clean drinking water from 130 km away, which is funded by UK charities and non-governmental organisations (NGOs).
- The scheme has many issues that limit its success, especially a lack of money for the locals to buy the water, little local technical know-how, and few spare parts for the machinery needed for the scheme to operate.

Dams help to increase water supply

Key Point

Conservation of water is an important strategy to save precious resources.

Quick Test

1. What strategies can be used to obtain water in areas where it is in short supply?
2. What methods of sustainable water management can be used in the home?
3. What are the advantages and disadvantages of using groundwater supplies?

Key Words

water transfer scheme
desalination
water conservation
aquifer
greywater

Energy 1

You must be able to:

- Identify areas of the world experiencing energy security and those suffering from energy insecurity or even energy poverty
- Explain how the exploitation of energy resources has many impacts on supply and the environment.

Energy Supply

- Demand for energy resources is rising globally but supply can be insecure, which may lead to conflict.
- Much of the world relies on **fossil fuels** that are non-renewable; reducing our reliance on these will increase our **energy security**.
- Some countries and regions are energy secure, while others suffer from **energy insecurity**.
- Eastern Europe, including Russia, has large reserves of natural gas and coal, with Russia among the top ten countries for oil and uranium production and thus energy secure.
- Much of the rest of Europe is heavily dependent on energy imports as it has declining fossil fuel supply and has energy insecurity.
- The Middle East and North Africa have large oil reserves but unstable governments often affect how much oil they can export.
- This is affected by the fluctuating price of crude oil or political factors such as the UN trade embargo on Iranian oil from 1995 to 2016.
- At present these countries are energy secure but diminishing reserves could change this status in the future.
- North America has large coal resources but its conventional oil resources are largely now exploited. It is now exploiting non-conventional oil and gas reserves in Alaska, the Alberta Oil Sands and through fracking, but its huge **energy consumption** often outweighs supplies and it is still, to some extent, energy insecure.
- Unconventional energy sources, such as fracking, often cost more to exploit than conventional sources due to the remoteness of the resource, security concerns and industrial processes needed to exploit the raw materials (e.g. oil sands).
- Asia (excluding Russia) has large coal and uranium reserves but has a rapidly increasing demand for oil in particular, outweighing available supplies. The region suffers from energy insecurity.
- Some African countries depend on foreign TNCs to exploit resources.
- Many African people use wood as their primary source of energy and much of the population suffers from **energy poverty**.

There are an estimated 1074 billion tonnes of proven coal reserves worldwide – enough to last about 130 years at current rates of production

Key Point

Energy supply is determined by many external factors, both physical and human.

- Energy consumption is increasing throughout the world for a number of reasons:
  - Economic development leads to increased consumption.
  - As rising wealth leads to increased living standards, this in turn leads to more electrical items and technology being used in the home and increased rates of car ownership.
  - As the world's population rises, so does the demand for energy.

Energy Usage

- The impacts of energy insecurity lead to energy supply problems for some regions.
- The variation in fossil fuel prices, especially oil and gas, may mean that the cost of exploitation and production in areas that are challenging will lead to shortages in the supply of these fuels.
- The development of technology has enabled humans to exploit energy sources in remote, difficult and environmentally sensitive areas, such as in deep oceans and polar regions.
- Political factors also affect the supply of energy, such as the conflicts in Middle Eastern countries in the late 20th and early 21st centuries (e.g. Iraq and Syria are heavily involved in the control of oilfields).
- In the 1960s, British companies searched for oil and gas in rocks underneath the North Sea, leading to Britain becoming self-sufficient in oil and gas for a while.
- In Alaska, exploitation of oil sources has taken place in environmentally sensitive and remote areas close to the Arctic Ocean.
- Agriculture uses oil products to power farm machinery, for the transport of goods and livestock, and in agricultural chemicals such as fertilisers and pesticides.
- In recent years, agricultural goods like barley, maize and sugar cane have been used to make **biofuels** that are used as a substitute for oil-based fuels.
- Rising prices for oil mean a higher price for biofuels and agricultural chemicals, making food more expensive.
- Oil is also used as a raw material in many industries such as plastics, packaging and textiles, so rises in the price of oil also lead to a higher price for goods.

Saudi Arabia, USA and Russia are the biggest oil producers

Key Point

Many countries struggle to obtain energy on a reliable basis at a price that can be sustained.

Quick Test

1. What regions of the world are energy secure and why?
2. Suggest how energy consumption rates vary in different climates.
3. How does the price of oil affect food prices?

Key Words

fossil fuel
energy security
energy insecurity
energy consumption
energy poverty
biofuel

Energy 2

Quick Recall Quiz

You must be able to:

- Describe strategies to increase energy supply
- Show how the extraction of a fossil fuel has both advantages and disadvantages.

Strategies to Increase Energy Supply

- Renewable sources of energy are more environmentally friendly than non-renewable sources of energy, but can be more expensive to generate electricity as the set up costs can be greater.
- Fossil fuels, especially coal and gas, are much more efficient at producing energy than renewables.
- All fossil fuels produce carbon dioxide (CO_2), which contributes to the enhanced greenhouse effect that causes climate change; and some produce other pollutants such as sulphur.
- Many countries, such as the UK and Germany, have used gas to replace coal as a source of energy to make electricity, because gas produces only half as much CO_2 as coal.
- **Clean coal technology** has enabled coal-fired power stations to produce less pollution and CO_2 by removing the pollutants when the coal is burned. However, in the UK they will all close by 2025.
- Nuclear power is seen by many to be the solution to electricity generation as it is a largely carbon-free source of energy.
- Some countries, such as Japan, are reducing their nuclear programmes due to environmental concerns.
- Other countries, such as the UK, are looking to increase their number of nuclear power stations to reduce greenhouse gas emissions.
- Some countries have large numbers of nuclear reactors that produce much of their electrical energy. For example, France has 56 reactors producing 71% of its electricity.

Renewable Energy Sources

- There are many strategies to increase energy supply not only by using non-renewable sources, but also by using renewable sources.
- **Biomass** is the use of plants to produce energy; this can include the burning of wood or the conversion of crops into biofuels such as ethanol.
- Many power stations in the UK have been converted to burn biomass as well as coal.

> **Key Point**
>
> Oil is the most important source of energy in the world at present – its impact stretches far beyond transport.

Cooling towers of a nuclear power plant

> **Key Point**
>
> Gas and nuclear power are replacing coal in many countries as a source of energy because they produce fewer greenhouse gases.

- Drax power station in South Yorkshire, originally built to burn coal from the nearby Selby coalfield (now closed), now burns wood pellets and processed straw.
- Selby was the UK's most recently opened coalfield in the 1970s, when coal was used to produce electrical energy in nearby power stations as the price of oil was rapidly rising.
- As coal was identified as a major source of carbon dioxide and workable seams became exhausted, production fell and Selby began to make losses, resulting in its closure in 2004.
- Wind power is the most widely used renewable in the UK, with more than 11 000 onshore and offshore wind turbines producing about one quarter of the UK's electricity.
- The Muppandal wind farm in Tamil Nadu in southern India has over 3000 wind turbines and is one of the largest in the world. It produces 1500 MW – about 20% of India's total wind power production.
- As India is a heavy user of coal, installations such as Muppandal reduce the amount of carbon dioxide and pollution produced, as well as providing jobs in what was a very poor area.
- Hydroelectric power (HEP) is the most established of the renewables as massive dams in many parts of the world, such as the Three Gorges Dam in China, produce vast amounts of electricity. However, micro HEP schemes now provide electricity in remote rural regions around the world, e.g. Bawan Valley on the island of Borneo.
- Tidal and wave power are forms of HEP that use the sea to produce electricity. Most tidal schemes use barrages that trap the incoming tide and release it when electricity demand is high to drive turbines and produce power.
- A tidal power station has been operational since 2008 in Strangford Lough in Northern Ireland.
- **Geothermal energy** is found mainly in areas where there is volcanic activity (such as Iceland) and the steam produced is used to power turbines to create electricity.
- Solar power has massive potential in countries with high annual amounts of sunshine. Even in the UK, with approximately 1 million solar power installations, up to 5% of total electricity generation is possible given the right conditions.

Wind turbines and solar panels

A geothermal power plant

Key Point

Renewable sources of energy are expensive to set up and do not provide as much energy as conventional sources.

Key Words

clean coal technology
biomass
geothermal energy

Energy 3

You must be able to:

- Explain why sustainable energy supplies are important for the survival of the planet
- Describe how energy usage by individuals and organisations can be designed to be sustainable
- Show how technology can be used to reduce energy usage.

Quick Recall Quiz

How Can Energy Usage be Made More Sustainable?

- A **carbon footprint** is the greenhouse gas emissions created by an organisation, event, product or individual.
- Carbon footprints in HICs tend to be high because of the lifestyles people lead, with a reliance on fossil fuels, heavy use of electrical devices at work and home, and a diet that relies on food that is imported or uses high levels of inputs derived from fossil fuels.
- In LICs, carbon footprints tend to be low because lifestyles are more sustainable as less fossil fuel is used and food tends to be locally produced.
- Energy conservation is important in protecting precious supplies of energy.
- Energy can be conserved through the intelligent design and building of homes and other buildings like offices and factories. Methods include the use of insulation in loft spaces and walls, double glazed windows and larger windows in south-facing walls.
- Using energy-efficient devices, like kettles that only boil one cup of water, or turning off appliances when not in use (rather than using standby buttons) can also help conserve energy in the home.
- The use of LED light bulbs, washing machines that wash clothes at 30°C or lower and thermostats on central heating systems that can be controlled using a mobile phone, are all ways in which homes and workplaces can be made more energy efficient.
- Transport in cities can be made more sustainable by encouraging people not to use cars. Public transport or bicycles are more sustainable and help to reduce carbon emissions.
- Examples of sustainable public transport measures include the Congestion Charge in central London, where cars are charged for entering the city centre to reduce the amount of traffic and cut down on pollution.
- Most of Greater London is covered by the Low Emission Zone, which charges diesel-powered commercial vehicles that do not meet certain emission standards.
- The introduction of bike lanes and bus lanes can help to speed up public transport in towns and cities.
- Park-and-ride schemes in a number of cities (like Oxford and Cambridge) have helped to reduce traffic in the centres.

Key Point

Carbon footprints tend to be higher in more developed countries owing to less sustainable lifestyles.

Congestion charging in London

- The re-introduction of trams into city centres like Manchester, Birmingham and Sheffield have helped to ease congestion in these cities and reduce pollution from diesel-powered buses.
- **Hybrid** buses have been introduced in many cities like London, Reading and Brighton, and they combine a conventional engine with an electric one, which results in greater energy efficiency and lower emissions.
- In many UK cities, the provision of bikes to hire at stations and bus stops has been popular following the example of London, which introduced so-called 'Boris Bikes' (named after the then Mayor, Boris Johnson, in 2010). There are now more than 11 000 of these bikes covering the central part of London and other cities like Liverpool have followed suit by introducing bikes for hire.
- Technology has been introduced to increase efficiency in the use of fossil fuels. This includes power stations where clean coal technology has been introduced.
- Cars too have become more energy efficient, with the introduction of new engines that burn fuel more efficiently, hybrid engines in cars like the Toyota Prius and more aerodynamic designs.
- The UK uses about the same amount of oil for transport as it did in the 1970s despite the number of vehicles doubling since then.
- Airliners too have seen similar levels of efficiency with the introduction of aircraft like the Boeing 787 Dreamliner. The Dreamliner uses 25% less fuel than conventional airliners like the Airbus 340 due to efficient engines and the use of lighter materials in the aircraft's construction.
- Education is important in making the public aware of energy conservation.
- Changing people's behaviour (or **adaptation**) in the use of energy is an effective way of reducing usage.
- Examples of energy conservation promotion include the use of posters beside light switches and the teaching of the topic in GCSE courses.

Key Point

It is possible to reduce energy usage through both the actions of individuals and organisations.

Dreamliner aircraft

Quick Test

1. What is a carbon footprint?
2. Suggest ways in which offices may be made more energy efficient.
3. In what ways can urban areas make transport more sustainable?
4. How has technology been used to make transport more energy efficient?

Key Words

carbon footprint
hybrid vehicle
adaptation

Fieldwork

You must be able to:

* Answer questions in an examination showing that you have made links between two fieldwork enquiries about the physical and human geography you have studied.

What is the Geographical Enquiry?

* A **geographical enquiry** is an investigation linking your study in class with fieldwork carried out away from school.
* You will have completed two geographical enquiries in two different places, studying the physical and human geography at these locations.

What Sort of Questions Can I Expect in the Examination?

Questions About How You Might Use Your Fieldwork Techniques in Unfamiliar Contexts

* You may be presented with new information and data about an investigation.
* What techniques were used or would you suggest using for collecting such data?
* What methods were used or would you suggest for presenting the data?
* How effective would these techniques be?

Students measuring orientation of glacial deposits in Snowdonia

Questions About Your Own Enquiries

* You will need to know the titles and be able to describe the location for both of your enquiries.
* Why did you choose the location for your enquiries?
  - Explain why the locations were suitable.
* Why did you choose your question?
  - How is it linked to geographical theories or concepts?
  - Explain any **hypothesis** or hypotheses you tested. A hypothesis is an idea or explanation for something that has not yet been proved. It could be described as an 'educated guess' about the possible outcome of your investigation.
* How did you prepare for your fieldwork?
  - In particular, what risk assessment did you conduct (an analysis of possible dangers) for both enquiries and what measures did you take to reduce these risks?

Field sketching in Carding Mill Valley, Shropshire

- Why did you collect your data?
  - You will need to describe and explain the **primary data** (data you and your fellow students have collected) and the **secondary data** (data collected by another person, group or organisation that you have accessed via articles, websites, books, etc.).
- Is your data **quantitative** (statistics; numbers)?
  - For example, width and depth of a river channel; an environmental quality survey.
- Or is your data **qualitative** (non-numerical; might involve subjective judgements)?
  - For example, field sketches; photographs; video; quotes or opinions.
- How did you collect your data?
- What methods of data collection did you use?
  - Mention **sampling techniques** you used to ensure the reliability of your data, e.g. random; systematic; stratified. If you used **GIS**, explain why such georeferenced data is useful.
  - Justify a method of data collection that you used.
- How did you process and present your data?
- Evaluate a method of data presentation that you used.
  - Describe and explain what you did with your data to make sense of it; what visual, graphical or cartographic methods (including GIS) you used and why.
- What did your data tell you?
  - Describe, analyse and explain your data.
  - What links did you find between data sets?
  - Which statistical techniques did you use?
  - What patterns and anomalies did you find?
- What conclusions did you make?
- Did your conclusions match your expectations at the beginning of the enquiry?
  - Make sure you refer to the original aims or hypotheses. A hypothesis may be correct or incorrect – it is important that you can explain why.
- How effective was your investigation?
  - How could you improve your data collection techniques?
  - What were the **limitations** of your data?
  - How reliable were your conclusions?

A student measuring river channel gradient in Carding Mill Valley, Shropshire

 Key Point

For both of your enquiries, you need to be able to describe and explain what you did at each stage and why. You must also be able to reach plausible conclusions supported by your data, evaluate your work and recognise any limitations.

 Key Words

geographical enquiry
hypothesis
primary data
secondary data
quantitative data
qualitative data
sampling techniques
GIS
limitation

Issue Evaluation

You must be able to:

- Apply geographical knowledge, understanding and skills to a particular issue.

What is the Issue Evaluation?

- The issue evaluation encourages **synoptic thinking**, which means applying the knowledge, understanding and skills from all of your geographical learning to a new situation at a range of scales from local, regional, national to international.
- The topic for the issue evaluation will be linked mainly to one of the compulsory (core) topics, but there will also be links to other topics.
- 12 weeks before the exam, a 'pre-release' resource booklet will be published about the issue.
- You can make notes on your resource booklet, but you cannot take this copy into the exam – a new copy will be provided for candidates.
- You will need to show that you understand the issue by referring to the resources.
- You will need to demonstrate **critical thinking** about the issue, which means showing your ability to interpret, analyse and evaluate ideas and arguments.
- **Problem-solving** is required: you will need to make a decision by choosing a possible option and justifying your decision.
- Longer, extended written answers will be expected for some questions.

> ## Key Point
>
> The issue evaluation encourages synoptic thinking. You will have to draw upon knowledge from all of the geography you have studied.

What Geographical Evidence is in the Resource Booklet?

- The pre-release booklet will contain a range of resources, such as different kinds of maps of different scales, GIS data, charts, diagrams, tables of statistics, photographs, aerial or satellite imagery, field sketches and written materials (e.g. web or newspaper articles; quotes from **stakeholders**, **key players** and **interest groups**).
- All the resources are important. You will need to use evidence from them to demonstrate your understanding of the issue and to support the arguments for or against a point of view. It may well be useful to use more than one resource at a time.

> ## Key Point
>
> You will need to take into account a wide range of facts and views to inform your decision-making.

How Can I Reach a Decision?

- Consider these questions:
  - What are the **impacts**? Remember, impacts can be positive as well as negative.
  - What are the main sources of conflict and agreement, e.g. between stakeholders, key players and interest groups.
  - How sustainable is each option or strategy?
  - How can I structure my thinking? Use categories such as advantages versus disadvantages; costs and benefits; SWOT analysis (strengths, weaknesses, opportunities, threats); economic, social and environmental considerations (remember that there are different types of pollution – air, water, noise, light, litter, etc.).
  - Why is **management** important? There are different approaches to governance, e.g. top-down (i.e. decisions are made from the top, such as national government, usually for large-scale schemes) versus bottom-up (i.e. decisions are based on everyone's opinion, such as those of local people, usually for small-scale schemes). Sustainable decisions consult people so they feel empowered (i.e. they feel they have greater ownership of the project and are therefore prepared to invest time and energy in supporting it).
  - How should I reach a decision? Explain reasons for choosing one option and reasons for not choosing other option(s). There will not be a 'right' or 'wrong' answer, but whatever decision you make must be supported by evidence from the resources.

Key Point

There is no 'correct' or 'incorrect' answer, but your decision must be supported by evidence in the resources.

Quick Test

1. What is meant by 'synoptic thinking' and 'critical thinking'?
2. Give definitions, with examples, of:
   a) stakeholders
   b) key players
   c) interest groups.
3. Explain what is meant by 'top-down' and 'bottom-up' approaches to management.
4. What is meant by 'empowerment'?

Key Words

synoptic thinking
critical thinking
problem-solving
stakeholder
key player
interest group
impact
management

Review Questions

Urbanisation

1 What is urbanisation? [1]

2 Identify two reasons why LIC cities are growing so fast. [2]

3 Which of the following statements are true? Which are false?

    a) In 1900, 40% of the world's population lived in urban areas.
    b) A megacity has a population of over 10 million people.
    c) Low wages are a 'pull factor' of rural-to-urban migration.
    d) The majority of the world's population now live in urban areas.
    e) The fastest rates of urban growth are experienced in higher income countries. [5]

4 Decide if each of these refer to **push factors** or **pull factors** of rural-to-urban migration:

    a) Unemployment   **b)** Higher wages
    c) Isolation   **d)** Unprofitable farming
    e) Better schools and hospitals [5]

Total Marks _____ / 13

Urban Issues and Challenges 1

1 Describe in detail two ways that quality of life can be improved in favelas such as Rocinha. [4]

2 Identify three positive aspects of life in Rio de Janeiro's favelas. [3]

3 Describe one environmental challenge faced by Rio de Janeiro. [2]

4 Describe the living conditions in Rio's favelas. [4]

Total Marks _____ / 13

Urban Issues and Challenges 2

1 Describe two environmental challenges faced by London. [4]

2 Describe two transport developments that have taken place in and around London. [4]

3 Give three reasons why tourists are attracted to London. [3]

4 Identify three social challenges faced by London. [3]

Total Marks _____ / 14

Urban Issues and Challenges 3

1 Cities can sometimes be described as 'systems'.

Identify two outputs of a city system. [2]

2 Identify two alternative fuel sources that could provide cleaner energy for transport systems. [2]

3 Explain ways that housing can be designed to be more sustainable. [4]

4 Describe two environmental improvements to urban areas that can form features
of sustainable living. [4]

Total Marks _____ / 12

Measuring Development and Quality of Life

1 What components make up the Human Development Index (HDI) and why has it become
popular as a measure of world development? [3]

2 Explain how a NEE (newly emerging economy) differs from a LIC (lower income country). [2]

3 Identify three population characteristics of LICs. [3]

Total Marks _____ / 8

The Development Gap

1 What historic causes have contributed to global inequalities? [2]

2 How can intermediate technology help to reduce the development gap? [4]

3 What is the difference between 'debt reduction' and 'debt relief'? [4]

4 What disadvantages might Kenya experience as it attempts to develop its
tourist industry? [4]

Total Marks _____ / 14

Review Questions

Changing Economic World Case Study – Vietnam

1 Outline the advantages and disadvantages of TNCs locating in countries with newly emerging economies. [6]

2 Describe how aid can bring benefits to an area. [3]

3 Discuss the impacts of economic development on people and the environment. [5]

Total Marks / 14

UK Economic Change

1 Which of these statements describing the UK north–south divide are true? Which are false?

a) Wages are higher in the north.
b) The north has traditionally depended on heavy industries.
c) People have longer life expectancy in the south.
d) Average house prices are higher in the north.
e) Unemployment rates are higher in the south. [5]

2 Explain why deindustrialisation has occurred in the UK. [3]

3 Discuss the social and economic changes in a rural landscape that has experienced population growth. [5]

Total Marks / 13

UK Economic Development

1 Name the countries that make up the G7. [3]

2 What major transport infrastructure improvements are planned for the UK in future years? [3]

3 Discuss how the UK is linked to the wider world. [5]

Total Marks / 11

Overview of Resources – UK

1. What are 'food miles'? [2]

2. Why is water use increasing in HICs like the UK? [3]

3. What strategies are employed to supply water to regions like south-east England, where demand is greater than supply? [3]

4. Explain why the development of hydraulic fracturing (fracking) potentially has both advantages and disadvantages for the UK. [5]

Total Marks _____ / 13

Food 1

1. What is meant by 'famine'? [1]

2. Explain how LICs can improve food security in their countries. [3]

3. List some of the physical and human factors that affect food production. [4]

4. Describe the potential consequences of climate change on food production for the poorest LICs. [4]

Total Marks _____ / 12

Food 2

1. What is meant by the term 'irrigation'? [1]

2. What is meant by the 'Blue Revolution'? [2]

3. Explain how any **two** of the following are able to increase food supply:

   hydroponics   **aeroponics**   **biotechnology**   **appropriate technology** [4]

4. For a large-scale irrigation scheme you have studied, describe how its development has both advantages and disadvantages. [5]

Total Marks _____ / 12

Practice Questions

Water 1

1 Define the term 'water insecurity'. [2]

2 Suggest two causes of water insecurity. [2]

3 As countries become more economically developed, their use of water increases.
Explain why this happens. [4]

Total Marks / 8

Water 2

1 What is meant by a 'water transfer scheme'? [1]

2 What is meant by 'desalination'? [1]

3 Describe the main features of a water transfer scheme you have studied. [5]

4 Why is groundwater an important water source for many countries? [5]

Total Marks / 12

Energy 1

Study the pie charts below. They show the sources of energy used to make electricity in the UK in 1970 and 2020.

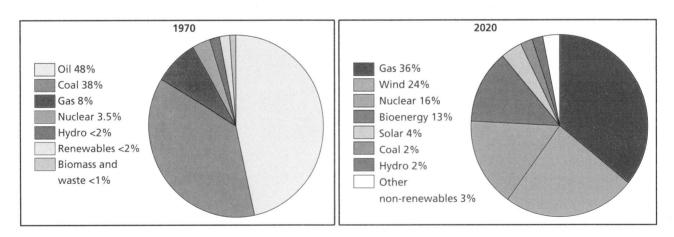

1970	2020
Oil 48%	Gas 36%
Coal 38%	Wind 24%
Gas 8%	Nuclear 16%
Nuclear 3.5%	Bioenergy 13%
Hydro <2%	Solar 4%
Renewables <2%	Coal 2%
Biomass and waste <1%	Hydro 2%
	Other non-renewables 3%

1 Describe the changes in the use of fossil fuels. [3]

2 **a)** What percentage of energy was made up of renewables in 2020? [1]
b) Why has this percentage increased since 1970? [3]

3 Why are regions like North America and western Europe energy insecure? [4]

Total Marks / 11

Energy 2

1. Describe what is meant by 'renewable energy sources'. [2]

2. Why are the owners of coal-powered electricity generating stations having to introduce new technology such as clean coal technology? [3]

3. Describe the main features of a renewable energy scheme that you have studied in a LIC or NEE to provide sustainable supplies of energy. [5]

4. Give some of the advantages and disadvantages of nuclear power. [8]

Total Marks _____ / 18

Energy 3

1. Why are carbon footprints high in HICs? [4]

2. Why are carbon footprints low in LICs? [2]

3. How can homes be made more energy efficient? [2]

4. What actions can people take in their homes to cut down on energy usage? [4]

Total Marks _____ / 12

Fieldwork

1. Suggest two data collection techniques that could be used to carry out a geographical fieldwork investigation in:

   a) a physical environment [1]
   b) a human environment. [1]

2. Explain some advantages of the locations for each of your fieldwork enquiries. [2]

3. Justify one primary data collection method used in relation to the aim(s) of your physical/human geography enquiry. [3]

4. Explain how a data collection technique could be improved to make the sample more reliable. [3]

Total Marks _____ / 10

Review Questions

Overview of Resources – UK

1. What is a 'carbon footprint'? [2]

2. Why does the UK have energy insecurity? [2]

3. Why is the development of hydraulic fracturing (fracking) likely to cause conflicts between different groups of people? [3]

4. How has the way that the UK uses energy changed in the last 50 years? [4]

Total Marks _____ / 11

Food 1

1. Define 'food security'. [2]

2. Suggest two physical causes of food insecurity. [2]

3. Suggest two human causes of food insecurity. [2]

4. Why do some parts of the African continent have food shortages? [4]

Total Marks _____ / 10

Food 2

1. Explain what is meant by 'organic farming'. [2]

2. How can reducing food waste in our homes enable us to be more sustainable? [2]

3. Describe how the 'Green Revolution' was able to increase food production in LICs. [4]

4. Describe the main features of a scheme in a LIC or NEE to increase sustainable supplies of food. [5]

Total Marks _____ / 13

Water 1

1. Explain the term 'physical water scarcity'. [1]

2. Explain the term 'economic water scarcity'. [2]

3. Explain how LICs can improve water security in their countries. [2]

4. Why will climate change have an impact on water security for many countries? [3]

Total Marks _____ / 8

Water 2

1. What is 'groundwater'? [1]
2. Suggest ways that water supplies can be increased in areas where water is scarce. [3]
3. Explain why the use of groundwater from aquifers is unsustainable. [3]
4. Describe the main features of a scheme that you have studied in a LIC or NEE to increase sustainable supplies of water. [5]

Total Marks .... / 12

Energy 1

1. Define 'energy security'. [1]
2. Suggest two causes of energy insecurity. [2]
3. Explain why the rising price of oil leads to higher food prices. [5]
4. Suggest reasons why oil resources are being exploited in remote, difficult and environmentally sensitive areas. [5]

Total Marks .... / 13

Energy 2

1. Define the term 'biofuel'. [1]
2. Why are some countries not suited for electricity production from HEP? [3]
3. What is 'geothermal energy'? [3]
4. What are the advantages and disadvantages of wind power for a HIC such as the UK? [4]

Total Marks .... / 11

Energy 3

1. What is 'energy conservation'? [2]
2. Describe how the design of buildings can be used to conserve energy. [3]
3. How can transport in cities be made more sustainable? [5]
4. What is 'clean coal technology'? [2]

Total Marks .... / 12

Fieldwork

1 Describe a data presentation technique you used for your primary or secondary data. [3]

2 Explain why you chose a particular data presentation technique for your fieldwork enquiry. [4]

3 From the photograph, identify the potential hazards and who may be affected by them. [4]

4 Outlining some of your risk assessment, how would you reduce the risks posed by the hazards in your fieldwork? [4]

Total Marks .... / 15

Mixed Questions

1 Study the figure below showing water usage per person in selected countries.

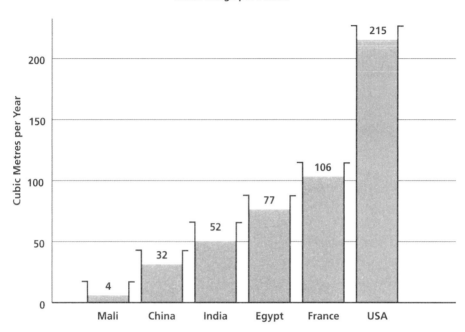

Water Usage per Person

Give reasons for the pattern of water usage shown in the graph. [5]

2 What is meant by a 'natural hazard'? [2]

3 State two examples of primary impacts of an earthquake. [2]

Mixed Questions

4 With reference to the Eyjafjallajökull volcanic eruption in 2010, which of the following statements are true and which are false? [5]

a) It occurred at a destructive plate margin.

b) It produced a giant ash plume.

c) It resulted in serious flooding.

d) It caused the cancellation of numerous air flights across Europe.

e) It occurred beneath an ice cap.

5 Describe how tropical storm intensity is measured on a scale. [2]

6 How does latitude influence the UK climate? [2]

7 The chart below is often referred to as the _____ graph. [1]

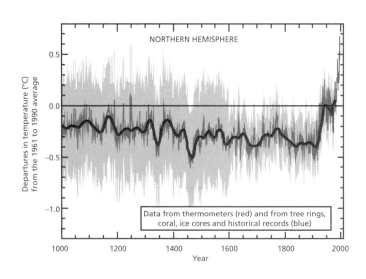

8 Describe one way in which an ecosystem naturally manages itself. [2]

9 Why are beaches said to be temporary features? [3]

10 What alternative name is often given to northern coniferous boreal forests? [1]

11 What problem can be caused by over-watering soil in desert areas? [1]

12 Give one reason which made the Lower Lea Valley in London difficult to regenerate. [1]

13 Apart from CITES and ecotourism, give two other ways in which rainforests can be sustainably managed. [2]

14 What are the causes and potential impacts of a storm surge with a tropical storm? [4]

15 Why have many terraced fields been abandoned? [1]

Mixed Questions

16 Give an alternative term for freeze-thaw action. [1]

17 How can the landscape influence transport and communication in upland glaciated areas? [2]

18 State two examples of secondary impacts of an earthquake. [2]

19 How can a family use energy more efficiently in the home? [4]

20 Suggest ways in which water resources can be conserved in HICs. [3]

21 Why is the most up-to-date technology not always the best solution to development problems in LICs? [2]

22 Explain the action of abrasion and how it changes a river channel. [3]

23 Diagram A shows the structure of rocks on a cliff.

Diagram B shows what the cliff looks like today.

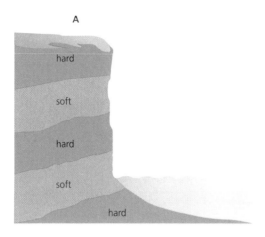

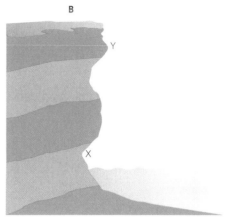

a) What feature is shown at X? [1]

b) Describe one physical weathering process that might affect area Y. [2]

c) What type of mass movement is most likely on this cliff? [1]

24 How does calorie intake vary between HICs and LICs? [3]

25 What is the difference between a 'greenfield' and a 'brownfield' development site? [2]

26 Why can waterfalls occur along the side of a U-shaped valley? [2]

Mixed Questions

27 On a separate sheet of paper, describe the formation of meanders using only labelled and/or annotated diagrams. [4]

28 In 1910, 10% of the world's population lived in urban areas.

What is the UN's predicted figure for 2050? [1]

29 Many local residents opposed the regeneration plans for Lower Lea Valley in London.

What might have been the reasons for this? [3]

30 Which three factors are combined to form the Human Development Index (or HDI) to measure development in different countries? [2]

31 What are the solutions that could end the UK's reliance on fossil fuels? [2]

32 How does ethical consumerism make agriculture more sustainable, especially in HICs? [2]

33 Explain the development of an arch from a cave. Refer to named processes. [4]

34 Describe the differences in carbon footprints in higher income countries compared to lower income countries. [6]

35 For one of your geography enquiries, to what extent were the results of this enquiry helpful in reaching a reliable conclusion(s)? [9 + 3 SPaG]

Total Marks .... / 98

Answers

Page 9 Quick Test
1. Destructive boundary – two plates travel towards each other and collide, with the denser plate sinking below the other; Constructive boundary – two plates pull apart to create new land.
2. **Any suitable answer, e.g.** They have always lived there; jobs; confidence in government to 'fix' things; 'it won't happen to me' attitude; fertile soils; valuable minerals; geothermal energy; tourism.

Page 11 Quick Test
1. Primary effects occur immediately while secondary effects occur later on, bringing more problems to those affected.
2. The focus is the point underground where the earthquake originates, while the epicentre is the point on the surface directly above the focus.

Page 13 Quick Test
1. Shield volcano
2. Pyroclastic flow – torrent of hot ash, rocks, gases and steam, moving at up to 450 mph; Lahar – 60 mph mudslide of melted snow and volcanic ash.

Page 15 Quick Test
1. Northern
2. Pacific Ocean
3. It can cause coastal flooding, which may lead to more casualties than the high winds.
4. They lose energy supply (warm water) and slow due to friction (especially if the land is hilly).
5. Frequency is how often something happens; intensity is strength and concentration.

Page 17 Quick Test
1. The Philippines
2. Typhoon Haiyan was a Category 5 tropical storm.
3. Primary effects: Landslides occurred across the landscape; storm surges of 5–6 m on the islands of Leyte and Samar; Tacloban Airport terminal was destroyed; the entire first floor of the Tacloban City Convention Center, which was serving as an evacuation shelter, was submerged and many drowned. Secondary effects: economic effects included high losses due to businesses being damaged or closed, and development was halted; social effects included homelessness (1.9 million people), displacement (6 million people), bereavement, disease due to the lack of food, water, shelter and medication, education (schools closed); environmental effects included damage to ecosystems and loss of farmland.
4. Warm surface water in the western Pacific; climate change may have contributed.
5. Immediate responses: much of the central Philippines (Visayas) was placed under a state of national emergency; worldwide relief effort: aid valued at over $500 million.
   Long-term responses: the authorities adopted much more pro-active strategies with 'zero casualty' targets; they tackled the issue of inertia by offering incentives such as free bags of rice to persuade people to leave their homes and property behind.

Page 19 Quick Test
1. The prevailing winds are westerlies from the Atlantic Ocean, bringing relatively warm, moist air from the North Atlantic Drift (or Gulf Stream).
2. a) **Any suitable answer, e.g.** river flooding; sea flooding; winter storms; snow and ice; or drought (e.g. the 2010–12 drought).
   b) **Any suitable answers, e.g. for 2010-12 drought:** Economic and social impacts: farmers struggled to provide water for livestock and harvest crops; low reservoir levels and hosepipe bans affecting six million consumers.
   Environmental impacts: groundwater and river levels very low, affecting aquatic ecosystems; wildfires spread.
   Management strategies: hosepipe bans were introduced; water meters were installed to monitor usage; water companies fixed leaking pipes; water-saving devices were encouraged; education about water use; improved waste-water recycling; new reservoirs and pipelines considered; desalination plants considered.

Page 21 Quick Test
1. **Any suitable answers, e.g.** ice cores; marine sediment cores; pollen analysis

2. **Any suitable answers, e.g.**
   Physical causes: orbital changes (Milankovitch cycles); volcanic activity (volcanic emissions block sunlight); solar output (changes in the Sun's energy).
   Human causes: use of fossil fuels producing greenhouse gases (e.g. power generation), transportation; agriculture (e.g. methane from cattle, rice paddies); deforestation (e.g. tree removal reduces natural carbon sequestration); methane release from melting permafrost and ocean floors (e.g. due to anthropogenic global warming).

3.

Climate change response	NOT adaptation	NOT mitigation
Electric cars	✓	
Higher sea walls		✓
Tidal power	✓	
Wind farm	✓	
IPCC carbon reduction targets	✓	
Improving air-conditioning in houses		✓

Page 22: Tectonic Hazards 1
1. Convection currents [1] in the mantle [1].
2. **Diagram should include at least four of:** convergence, denser oceanic plate, less dense continental plate, melting plate, rising magma, fold mountains. For example:

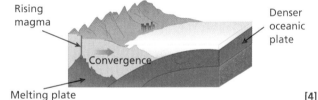

Rising magma; Denser oceanic plate; Convergence; Melting plate [4]

3. **Diagram should include at least four of:** divergence, rising magma, underwater volcanoes, new rock formed, spreading sea floor, oceanic ridge. For example:

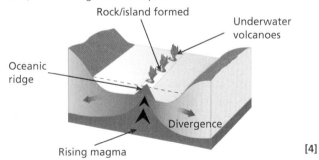

Rock/island formed; Underwater volcanoes; Oceanic ridge; Divergence; Rising magma [4]

4. **Diagram should include at least:** direction of movement, friction, faults, crust not destroyed. For example:

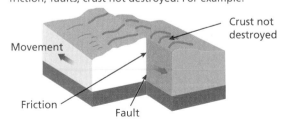

Crust not destroyed; Movement; Friction; Fault [4]

Page 22: Tectonic Hazards 2
1. LIC – lower income country [1]; HIC – higher income country [1]
2. A – True [1]; B – True [1]; C – True [1]; D – True [1]; E – False [1]
3. **Answer may include references to the factors of:** Degree of preparedness, HIC v LIC, population density, quality of buildings,

poverty, depth of focus, etc. [4], and include named examples (such as Japan 2011 – high magnitude and high death toll; Haiti 2010 – lower magnitude and high death toll). [2]

Page 22: Tectonic Hazards 3

1. **Any suitable answers with at least two primary effects and two secondary effects included:**
   Primary effects included: more than 350 people killed; 360 000 people displaced to emergency shelters; volcanic bombs and hot gases spread for 11 km; pyroclastic flows of up to 9 miles; ash fell up to 30 km away; nearby villages were buried. [2]
   Secondary effects included: 350 000 people left homes in the area; sulphur dioxide was blown across Indonesia and as far as Australia; ash cloud disrupted air travel; roads were blocked; lahars; food prices increased; airports closed. [2]

2. **Any two from:** Volcanic ash plume at 11 000 metres; fine-grained ash was a hazard to air traffic; lava flows; severe flooding; damage to roads, bridges, water supplies and livestock. [2]

3. **Any suitable answer, e.g.** Continued research into improving forecast methods for eruptions; improved warning and evacuation procedures; the construction of safer homes; people moved away from the most vulnerable areas around a volcano; further research into the effects of ash eruptions on air traffic; the construction of dams to hold back lahars; financial support made available to affected farmers and residents. [Up to 4 marks]

Page 23: Tropical Storms

1. B [1]
2. **Any three suitable features, e.g.** The tropical storm has a circular shape; the cloud is spinning in an anti-clockwise direction inwards towards the centre (which means it is in the Northern Hemisphere); there is a vortex; the storm has an eye. [3]
3. a) Cyclone [1]
   b) Hurricane [1]
   c) Typhoon [1]
4. The Coriolis force (Coriolis effect) makes low pressure systems spin anti-clockwise in the Northern Hemisphere [2] and clockwise in the Southern Hemisphere. [2]

Page 24: Tropical Storms – Case Study

1. Prediction [1]; protection [1]; planning [1]
2. **Any suitable example, e.g.**
   Name: Typhoon Haiyan [1]
   Where: The Philippines, south-east Asia (capital: Manila) [1]
   When: November 2013 [1]

3.

Impacts	Economic	Social/ Political	Environmental
Homelessness		✓	
Factories and other businesses closed or inaccessible due to damage to transport infrastructure	✓		
Waterborne diseases		✓	
Damage to ecosystems			✓
Schools closed for weeks		✓	

[1 for each correct row]

4. a) P [1]   b) S [1]   c) S [1]
   d) P [1]   e) P [1]   f) S [1]

Page 24: Extreme Weather in the UK

1. **Any suitable answers, e.g.** 2010–12 drought [1]
2. Higher ground in the west has greater precipitation (relief rainfall) and lower temperatures [1]. Consequently the east is in a rain shadow, receiving much lower precipitation. [1]
3. **Any suitable hydro-meteorological hazards linked to the water cycle, e.g.** River flooding; sea flooding; winter storms; snow and ice; drought [3]
4. **Any suitable answers, e.g.** Blocking highs (slow-moving anticyclones) led to very dry winters in 2009–10 and 2011–12 [1]. East winds were common, bringing in dry continental air [1]. Significantly lower precipitation than normal [1]. Climate change may have contributed. [1]

Page 25: Climate Change

1. D [1]

2. Anthropogenic factors [1]
3. **Any suitable causes, e.g.** Use of fossil fuels producing greenhouse gases (e.g. power generation; transportation); agriculture (e.g. methane from cattle; rice paddies); deforestation (e.g. tree removal reduces natural carbon sequestration); methane release from landfill sites. [3]
4. Interglacial – Warmer periods of glacier retreat
   Glacial – Extremely cold periods of glacier growth
   Proxy measure – Indirect ways to find out average temperatures from the past
   Quaternary – Last 2.6 million years, including the 'Ice Age'
   [3 if all correct; 2 if two correct; 1 if one correct]

Page 27 Quick Test

1. **Any suitable answers, e.g.**
   Large-scale: tropical rainforest, hot desert, temperate deciduous forest
   Small-scale: pond, hedgerow
2. Food chains show simple relationships between elements within an ecosystem, whereas food webs show complex interrelationships.
3. There will be more food for the prey of that animal.
4. Bird eats seed; bird dies; fungi break down remains; trees reabsorb nutrients from fungi.

Page 29 Quick Test

1. Along the Equator and between the Tropics of Cancer and Capricorn.
2. Mountains, coasts and rivers.
3. Western Europe, north-eastern USA, eastern China, Japan.

Page 31 Quick Test

1. Hot and wet all year round.
2. Trees (e.g. mahogany and kapok) have buttress roots; pitcher plants catch insects.
3. Hot and dry all year round.
4. Saguaro cactus has no leaves to cut down transpiration; quiver tree has fleshy leaves to store moisture.
5. **Any suitable answers, e.g.** Sahara; Kalahari; Mojave.

Page 33 Quick Test

1. **Any suitable answers, e.g.** Mining; settlement; roads; HEP.
2. **Any suitable answers, e.g.** Loss of species; loss of habitats; flooding of forest; soil erosion; locals lose land; reduced rainfall rates
3. **Any suitable answer, e.g.** Selective cutting / logging; ecotourism; labelling schemes; international agreements.

Page 35 Quick Test

1. **Any suitable answers, e.g.** Climate change; population growth; overgrazing.
2. Salts are deposited in the soil surface when water in soils evaporates in high temperatures.
3. **Any suitable answer, e.g.** Encouraging less water use among local people; developing planning laws that restrict the size of buildings; planting trees to stabilise sand dunes; encouraging the use of drip irrigation.

Page 37 Quick Test

1. Cold and dry in winter, mild and dry in summer.
2. Land that has been frozen for at least two consecutive years.
3. Benefits – jobs, profits; drawbacks – environmental destruction.
4. Tourists being attracted to new and undiscovered places.

Page 38: Tectonic Hazards 1

1. An area that develops above a mantle plume of rising heat from deep in the Earth [1]. Magma generated rises and works its way through a thin section of crust to the surface [1].
2. **Any suitable answers, e.g.** Poor quality housing; poor infrastructure making it harder to reach affected people; less money to protect people (e.g. earthquake-proof buildings); less money for responses (e.g. providing food and water); poor healthcare and facilities. [3]
3. A – Crust [1]; B – Mantle [1]; C – Inner core [1]; D – Outer core [1]
4. **Any suitable answers, e.g.** Building design; monitoring; firebreaks; training of emergency services; education of people in evacuation procedures; planned evacuations; survival kits; overseas aid [6]

Page 38: Tectonic Hazards 2

1. Richter scale [1]
2. **Any suitable answers with references to the differences between the following in HICs and LICs:** Number of deaths; number of homeless; building damage; effect of communications problems

on people's lives; quality of construction of housing; quality of infrastructure influences ease of access to affected areas; ability to provide healthcare; population density; need for overseas aid. [6]

Page 38: Tectonic Hazards 3

1. **Diagram should include at least four of:** Crater, gentle slopes, low wide cone, lava layers, runny lava. For example: [4]

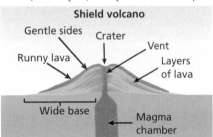

Shield volcano

Gentle sides
Crater
Vent
Runny lava
Layers of lava
Wide base
Magma chamber

2. Ash cloud – Blocks out the Sun, causing suffocation and health problems
Lahar – Mudslide including rock debris and water
Pyroclastic flow – Torrent of hot ash, rock, and gases and steam
Extinct volcano – Volcano nobody expects to erupt ever again
Dormant volcano – Volcano that has erupted in past 2000 years but is not currently active
[4 if all correct; 3 if three correct; 2 if two correct; 1 if one correct]

Page 38: Tropical Storms

1. In excess of 74 mph (119 km/h) [1]
2. The intense low pressure [1] of a tropical storm creates a 'dome' of seawater [1] and a storm surge occurs that causes coastal flooding. [1]
3. Warmer oceans expand, so storm surges may be worse [1]. Climate change may alter the distribution of tropical storms, and their frequency and intensity may increase [1], but the evidence for this is inconclusive [1].

4.

Country	Name used for tropical storms
Mexico	Hurricanes
Philippines	**Typhoons**
Australia	Cyclones
Bangladesh	**Cyclones**

[1 for each correct column]

Page 39: Tropical Storms – Case Study

1. B [1]
2. When people choose to stay in their homes [1], even though they have been warned that they may be under threat from a natural hazard [1].
3. Human factors can make matters worse (3 Ps). Prediction – high level warnings may be delayed (poor quality of governance and communications) [1]; Protection – communications infrastructure may be too vulnerable (e.g. it failed in the Visayas in the Philippines) and disease due to the lack of food, water, shelter and medication [1]; Planning – even though many people may be warned, they may choose to stay in their homes (inertia). [1]
4. Foreign investment in new infrastructure, e.g. more resilient bridges and railway – Long-term international response
Government declares state of emergency across the whole country – Immediate national response
Worldwide relief effort: aid valued at over $500 million – Immediate international response
Inertia tackled by offering incentives such as free bags of rice to encourage people to leave their homes – Long-term national response.
[3 if all correct; 2 two correct; 1 if one correct]

Page 40: Extreme Weather in the UK

1. **Any suitable answers, e.g.** Economic and social impacts: farmers struggled to provide water for livestock and harvest crops; low reservoir levels and hosepipe bans affected six million consumers. [1] Environmental impacts: groundwater and river levels were very low, affecting aquatic ecosystems; wildfires. [1]
2. **Any three from:** Hosepipe bans; water meters installed; water companies fix leaking pipes; water saving devices; education about water use; improved waste-water recycling; new reservoirs and pipelines considered; desalination plants considered (but high cost). [3]

3. The prevailing winds are westerlies from the Atlantic Ocean [1], creating a dominant maritime influence [1], bringing relatively warm, moist air from the North Atlantic Drift or Gulf Stream ocean current [1]
4. Place A – Mild winters, cool summers
Place B – Mild winters, warm summers
Place C – Cold winters, cool summers
Place D – Cold winters, warm summers
[3 if all correct; 2 if two correct; 1 if one correct]

Page 40: Climate Change

1. B [1]
2. Milankovitch cycles [1]
3. Volcanic eruptions can put enough gas and dust into the Earth's upper atmosphere [1] to block out solar energy, which can lead to a volcanic winter. [1]
4. Developing more drought-resistant crops or irrigation schemes – A [1]
Making buildings more energy efficient – M [1]
Higher flood defences along coasts and rivers – A [1]
Greater use of renewable resources – M [1]
Carbon capture and storage – M [1]
Planting more trees – M [1]

Page 41: Ecosystems and Balance

1. **Any suitable answer, e.g.** Pond; hedgerow. [1]
2. A community of biotic (living) [1] and abiotic (non-living) components that create an environment. [1]
3. Food chains show simple relationships between organisms [1]; food webs show more complex relationships. [1]
4. Plants grow and are eaten by herbivores [1]; the herbivores are then eaten by carnivores [1]; the carnivores die and decompose [1]; the decomposed remains are reabsorbed by the trees [1].

Page 41: Ecosystems and Global Atmospheric Circulation

1. High pressure [1]
2. The movement of air [1] owing to many factors, such as the movement of air masses. [1]
3. **Any two from:** mountains; coasts; rivers [2]
4. They are found in a wide belt encircling the Earth, roughly following the Equator and for the most part between the Tropics of Cancer and Capricorn [2]. The Amazon in the northern part of South America and the Congo basin in the centre-west of Africa are the largest rainforests. [2]

Page 42: Rainforests and Hot Deserts – Characteristics and Adaptations

1. Latosol [1]
2. They provide stability [1] and absorb nutrients directly from the fast-decaying leaf litter [1].
3. The climate in tropical rainforests has average daily temperatures of around 27–29°C [1]. Rainfall is high at around 500–600 mm each month [1].
4. Fennec foxes are nocturnal, which allows them to avoid the extreme temperatures of the day [1]. The fennec also has large ears [1], which act as radiators to lose heat [1].

Page 42: Opportunities, Threats and Management Strategies in the Amazon

1. **Any suitable answers, e.g.** iron; bauxite; copper; nickel; oil [2]
2. Only trees above a certain height are felled [1], thus leaving smaller trees to attain maturity [1].
3. **Any suitable answers, e.g.** Hunter-gatherers collect edible plants and catch wild animals to eat [1]; soils can be fertilised through small-scale shifting cultivation of crops such as manioc and cassava [1]; trees provide fuel wood and building materials [1]; medicines and hunting poisons can be extracted from a wide variety of plants and animals [1].
4. **Any two from:** Possible jobs for local people; money brought into the area; energy helps develop local businesses [2]

Page 43: Opportunities, Threats and Management Strategies in Hot Deserts

1. Infrastructure [1]
2. Hydroelectric power [1]
3. Sheep farming has led to the removal of vegetation cover [1] through overgrazing and trampling of the soil [1].
4. **Any suitable answers, e.g.** Encouraging less water use among local people [1]; developing planning laws that restrict the size of buildings [1]; planting trees to stabilise sand dunes [1]; encouraging the use of drip irrigation in farming [1].

Page 43: Polar and Tundra Environments
1. Mostly 60–70° north [1]
2. Extraction creates huge numbers of jobs [1] but destroys habitats [1].
3. Moderate summer temperatures of 20–25°C [1]. Winter temperatures can plunge below –40°C [1].
4. Plants such as the Arctic poppy [1] survive by developing adaptations such as shallow roots [1] and flowers that track the path of the Sun [1].

Page 45 Quick Test
1. Nearer the sea / away from the cliff
2. Rounded / low / slumped
3. The distance over which a wave has travelled
4. Plunging

Page 47 Quick Test
1. Steep
2. Distal end
3. Stump (or pediment)
4. The area between headlands / the eroded area between headlands.

Page 49 Quick Test
1. Longshore drift
2. To allow water to drain through gently / slowly
3. Hard
4. Reflected away (not absorbed)

Page 51 Quick Test
1. The load
2. It makes them smaller and rounder
3. It makes the channel wider
4. Anywhere along the course of the river

Page 53 Quick Test
1. Gorge
2. Slip-off slope
3. Moving water
4. Fine material deposited by rivers

Page 55 Quick Test
1. Cross-sectional area and velocity of water
2. The difference in minutes or hours between peak / highest rainfall and peak / highest discharge.
3. **Any suitable answer, e.g.** Tarmac; concrete; frozen ground; baked hard ground; some rocks, e.g. slate.
4. Water moving below the surface (e.g. in soil) towards the channel

Page 57 Quick Test
1. It happens as a result of exposure; no moving agent is involved.
2. Plucking
3. Lowland
4. Jagged / sharp edges

Page 59 Quick Test
1. Between two corries or along the back wall of corries.
2. Truncated spurs; hanging valley / waterfall.
3. Rugged / rocky / lumpy / irregular
4. Mixed sizes and types of rock fragments held in clay.

Page 61 Quick Test
1. Use of farmland or agricultural buildings for non-farming activities.
2. **Any suitable answer, e.g.** Climbing; fell-walking; kayaking; sailing; mountain biking; skiing.
3. **Any suitable answer, e.g.** Water supply to other areas; power supply for iron, textiles, paper production (water wheels or HEP); raw material (brewing); washing (textiles).
4. Sheep farming

Page 62: Ecosystems and Balance
1. **Any suitable answer, e.g.** Tropical rainforest; hot desert; temperate deciduous rainforest; boreal / taiga forest.
2. An organism that creates its own food [1] through photosynthesis [1].
3. An organism that eats other organisms [1] and can be herbivore or carnivore [1].
4. An organism that breaks down [1] the dead remains of other organisms [1].

Page 62: Ecosystems and Global Atmospheric Circulation
1. The Equator [1]
2. The Tropic of Cancer [1]
3. The Tropic of Capricorn [1]
4. They are found in a wide belt encircling the Earth in what are known as the high latitudes – areas on and a little south of the Arctic Circle [2]. Much of northern Russia, Canada and Iceland have tundra-type environments [2].

Page 62: Rainforests and Hot Deserts – Characteristics and Adaptations
1. Leaf litter [1]
2. Irrigation [1]
3. They attract insects using scent glands [1] and then catch them using a slippery flower, before digesting them [1].
4. Summer temperatures reach about 40°C and winter temperatures up to about 20°C [1]. Rainfall is low at 3 mm each month [1].

Page 63: Opportunities, Threats and Management Strategies in the Amazon
1. Convention on International Trade in Endangered Species [1]
2. Erosion [1]
3. Commercial means farming for profit [1], whereas subsistence means farming to feed oneself [1].
4. **Any suitable answer, e.g.** Schemes such as that at Tucuruí in Brazil [1] created huge amounts of energy for a growing economy [1] but also flooded huge areas of forest [1]. The dam also encouraged further deforestation for farming and settlement. [1]

Page 63: Opportunities, Threats and Management Strategies in Hot Deserts
1. Hoover Dam [1]
2. Gypsum [1]
3. Small areas of flat land created in hillsides [1], often in the path of natural watercourses [1].
4. Encouraging the use of drip irrigation [1]; implementing appropriate technologies that are cheap and easy to apply by local farmers [1]; recycling water within tourist areas [1]; controlling the size and number of golf courses [1].

Page 63: Polar and Tundra Environments
1. Permafrost [1]
2. Antarctic Treaty [1]; Madrid Protocol [1]
3. The Arctic fox [1] develops a thick coat to protect against the cold [1]. The Arctic hare has small ears to reduce heat loss [1] and white fur to avoid predators [1].
4. **Any three from**: Oil; bauxite; coal; fish [3]

Page 64: Coasts 1: Processes
1. Loose material is moved up and down the beach by waves [1]. Attrition / the particles crash together [1] so that over time, edges get rounded and size is reduced.
2. **Any suitable answer which gives a sense of the nature of the rock and cliff process, e.g.** More resistant rock gives steeper cliffs than softer rock. Resistant rock can support higher and steeper land. Softer rock is more easily weathered so the surface is lowered and slumping takes place, giving rounded cliffs. [4] **Or** More resistant rock is not easily weathered but particles break away and fall. The cliff moves back / retreats parallel to the original shape. [4]

Page 64: Coasts 2: Landforms
1. a) Cave [1]
   b) Arch [1]
2. Flow of water from a nearby estuary [1] could prevent further growth so that material is deposited. Second most frequent / dominant waves push the end of the spit round [1].
3. **Any three from:** Strong onshore winds / winds blowing most often from the sea; wide area of sand between high and low tide marks / sand that dries out between tides; mainly sandy beach rather than pebbles or shingle; growth of plants to stabilise the sand (lyme grass, sea couch or marram grass); gently sloping or flat area on landward side of the beach. [3]

Page 64: Coasts 3: Management
1. **Any suitable answer with a negative point stated and explained (no credit for description of walls unless used to amplify a point), e.g.** Sea walls tend to reflect wave energy rather than absorb it so there is still energy left to erode / scrape away material at the base of the wall or further down the beach. This makes the base of the wall unstable so likely to fail / need expensive repairs. [3] **Or** Sea walls are not attractive / are made of concrete / not made of material like the coast (or similar). The coast attracts people for recreation, especially on the beach but some walls make access to the beach difficult. To make them attractive, with a promenade and gardens on top, costs a lot of money. [3]
2. **Any suitable answer making specific comparisons, with emphasis on the techniques employed, their strengths and their weaknesses, e.g.** Nourishment is adding material to a beach that has suffered erosion. This material is usually imported with

risks of poor match of size and composition, which can create new problems. It is quite expensive. It usually improves the attractiveness of a beach to people **[2]**. Reprofiling uses local sand / material, redistributing it up the beach to change the shape of the beach so that it should be more resilient. It is usually cheaper. It may not make a beach more attractive to people. **[2]**

Page 65: Rivers 1: Processes
1. Carbonates, e.g. limestone **[1]**
2. **Any one from:** Saltation; traction **[1]**
   Any two points from: Saltation is particles / pieces of rock 'jumping' from the channel bed up **[1]** into the flow of the water then falling back down **[1]**. The impact of this can cause another particle to move **[1]**.
   Or any two points from: Traction is the dragging / rolling **[1]** or sliding **[1]** of material along the channel bed / floor or over other particles **[1]**.
3. **Any suitable answer which covers both ideas and shows understanding that not all eroded particles will be carried, e.g.** Entrainment is the taking in of material from the channel banks or bed **[1]**, which can then be used for abrasion **[1]** to further erode the channel **[1]**. Some eroded material may not be entrained **[1]**, and just falls to the channel bed. **[Up to 3 marks]**

Page 65: Rivers 2: Landforms
1. **a)** A – waterfall **[1]**; B – floodplain **[1]**
   b) Ox-bow lake **[1]**
   c) Source **[1]**
   d) Headward erosion / lengthening of the stream channel **[1]**
2. **Any suitable diagram, e.g.**

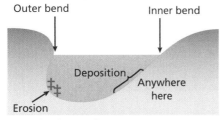

Outer bend · Inner bend · Deposition · Anywhere here · Erosion **[4]**

Page 66: Rivers 3: Flooding and Management
1. **Any two from:** A small catchment area; surrounded by steep slopes; has a lot of tributaries; surrounded by impermeable slopes **[2]**
2. **a)** 2 hours **[1 – unit must be given]**
   b) 50 mm **[1 – unit must be given]**
   c) 4.5 cumecs **[1 – unit must be given]**
   d) 9 **[1]**
3. **Any suitable answer which shows understanding of barriers to water getting into a channel or of those water movement systems that are slower, e.g.**
   Rainwater could be intercepted **[1]** by trees, buildings, etc., then drip to the floor **[1]**, infiltrate / pass into the soil / ground **[1]** and then move towards the channel as throughflow **[1]**.
   Or Rainwater could fall onto permeable ground **[1]**, then infiltrate / pass into the soil **[1]** and continue to travel vertically into the rocks **[1]**, where it will be held and eventually pass laterally / sideways as groundwater flow **[1]** to the channel.

Page 67: Glaciation 1: Processes
1. Terminal moraine **[1]**
2. At the base and sides **[1]** of a glacier where it meets the bedrock **[1]**
3. **Any suitable answer, e.g.**

Direction of ice flow

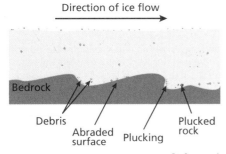

Bedrock · Debris · Abraded surface · Plucking · Plucked rock

[1 for each correct label]

Page 67: Glaciation 2: Landscape
1. At the beginning / inside edge of a trough / valley **[1]**
2. **Any two from:** Steep; rocky; frost-shattered **[2]**

3. Valley floors were made rugged by ice erosion **[1]**; plucking left hollows **[1]**; meltwaters washed debris into hollows **[1]** and rivers are still bringing material down **[1]** so that a flat valley floor is created and maintained.

Page 67: Glaciation 3: Land Use and Issues
1. **Any two ideas with appropriate explanation from:** High rainfall **[1]** because of altitude / relief rainfall **[1]**; steep slopes **[1]** mean water flows quickly towards valley floors **[1]**; impermeable rock **[1]** means that rainwater does not get absorbed / infiltrate **[1]**; flat valley floors **[1]** mean that the water does not flow away **[1]**.
2. **Any two features with connected activity from:** Steep, rocky truncated spurs **[1]** are good for climbing / bouldering **[1]**; ribbon lakes or corrie lakes **[1]** can be used for sailing or kayaking / canoeing **[1]**; high land gets more snow than elsewhere **[1]** so skiing can take place; mountain scenery attracts walkers **[1]**; arêtes make routes up to the high peaks for walkers **[1]**.

Pages 68–85 Revise Questions

Page 69 Quick Test
1. A city with a population of over 1 million people.
2. 56%
3. Over 10 million

Page 71 Quick Test
1. About 24%
2. **Any suitable answers, e.g.** Favela Bairro Project; upgrading housing; providing pavements; provision of electricity; new sewage systems; legal ownership rights; self-help schemes; low rents; improved transport systems.
3. Local residents are provided with materials like concrete blocks and cement to construct permanent buildings with water and sanitation.

Page 73 Quick Test
1. **Any suitable answers, e.g.** Loss of manufacturing industry; pay inequality; unemployment
2. **Any suitable answers, e.g.** The Crossrail project; HS2; new runway at Heathrow Airport
3. The home ground for West Ham United Football Club

Page 75 Quick Test
1. **Any suitable answers, e.g.** Renewable energy use such as solar panels; insulation; affordable prices and rents; passive energy use.
2. **Any suitable answers, e.g.** Car sharing; park-and-ride; vehicle restriction zones; integrated public transport systems; alternative fuels; cycleways; walkways.
3. A city which produces as much energy as is used (i.e. 'carbon neutral'); a city which supports sustainable lifestyles, such as renewable energy use, local services and jobs, locally-produced food.

Page 77 Quick Test
1. **Any suitable answers, e.g.** Life expectancy; adult literacy; birth rate; people per doctor, etc.
2. **Any suitable answers, e.g.** HICs – UK, France, Australia, USA; LICs – Ethiopia, Mozambique, Mali, Democratic Republic of Congo; NEEs – China, India, Brazil, Mexico, South Africa
3. Brazil, Russia, India, China and South Africa

Page 79 Quick Test
1. People move to improve their quality of life (this may be voluntarily or by being forced).
2. Long-term aid is trying to improve people's quality of life, such as access to water, healthcare or education, whereas short-term aid is often in response to a crisis such as providing shelter for people after a natural disaster.
3. Fair trade is where producers receive a guaranteed fair price for the things they make and grow.

Page 81 Quick Test
1. Cheap labour; a growing home market; fewer industrial laws and restrictions; lax environmental laws
2. Multilateral aid is given by countries through an international organisation like the World Bank, whereas bilateral aid is given from one country to another.
3. **Any suitable answers, e.g.** Deforestation; destruction of habitats; air pollution; water pollution; release of greenhouse gases; soil erosion; increase in waste.

Page 83 Quick Test
1. **Any suitable answers, e.g.** Mechanisation of primary industries; depletion of raw materials; competition from overseas; cheaper labour costs overseas; more advanced production methods overseas
2. An area where businesses receive incentives to locate, e.g. tax breaks

3. Any suitable answers, **e.g.** Economic – increased house prices; local shops and services are supported; new businesses set up; investment into housing improvements; social – schools oversubscribed; traffic congestion leading to longer journey times; services aimed at wealthier residents; community events

Page 85 Quick Test

1. Globalisation is the process by which the world is becoming increasingly interconnected.
2. Any suitable answers e.g. Reduced pressure on existing road and train networks; shorter journey times between major cities; creates jobs in the construction industry; encourages economic growth

Pages 86–88 Review Questions

Page 86: Coasts 1: Processes

1. Waves that add material to a beach **[1]**. More effective swash than backwash **[1]**. **[Allow 'stronger' swash or the reverse – weaker backwash]**
2. **Any two from:** Surging waves push material up the beach **[1]** so it will become steeper **[1]** and higher **[1]**. A ridge of pebbles / shingle / material might be left at the top / landward end of the beach **[1]**. **[Up to 2 marks]**
3. **Any suitable answer which describes hydraulic action, the importance of wave action and the results, e.g.** Hydraulic action is a process of erosion. It is very effective on rocks with faults or cracks or joints or visible bedding planes. Approaching waves trap air into the spaces. The air exerts great pressure against the back / sides of the space and loosens rock particles which then fall from the cliff. **[3]**

Page 86: Coasts 2: Landforms

1. **Any suitable answer which addresses the start of the process and gives a sense of the cliff disappearing but the 'floor' being left (no credit for description, only explanation), e.g.**
It starts with a notch formed by waves at high water / wave attack low on a cliff / at cliff front. This weakens the cliff / causes the cliff eventually to collapse and retreat at that level. The base of the cliff / underlying rocks are then exposed and eroded mainly by abrasion. **[3]**
2. **Any suitable answers with a set of ideas that show understanding of the apparent contradiction of weaker bays no longer being eroded, e.g.** Bays form from weaker areas between more resistant headlands but become sheltered by the headlands. As headlands stick out into the sea, wave attack is strong. Waves have further to travel into a bay and have lost energy after breaking on headlands / waves going into bays are low energy, which push sand into the bays / create beaches in the bays. Once a beach is established, waves have further to travel so do not erode the bay any more. **[4]**

Page 86: Coasts 3: Management

1. **Any suitable answer which addresses the beach as physical protection, e.g.** The further a wave has to travel after breaking, the less erosive power it has. Beaches absorb wave energy / do not reflect it. Material will be deposited, creating even more effective protection. **[3]** **[Up to 1 mark for any reference to the aesthetic value of beaches, their ecosystem value and economic role in tourism, as long as the point is linked to 'protection'.]**
2. **Any suitable answer, e.g.** Dunes are made of sand grains that are not consolidated / not stable **[1]**. Walking across / sliding down dunes dislodges the sand **[1]** and breaks down the shape of the dune **[1]** so that wind can blow through instead of bringing more sand to increase the dune size **[1]**. Vegetation on dunes is not a complete cover unless they are very old **[1]** and walking will easily break the stems / leaves / blades to stop the plant doing its job **[1]**. Dunes are very attractive, as they have sheltered sections out of the wind, but to reach them people must walk over the ridges **[1]**. **[Up to 3 marks]**
3. **Any suitable answer which shows the idea of change from... to... , e.g.** Before building, the farmland had a long, even width of beach in front **[1]** but after building the beach has partially disappeared **[1]**. Sand / material has piled up against the updrift side / western side of the jetty **[1]** but has been eroded / moved away from the jetty / on downdrift side / to the east of the jetty **[1]**, putting the town at risk **[1]**. **[Up to 4 marks]**

Page 87: Rivers 1: Processes

1. Sand is coarser and so does not lock / stick together **[1]** as well as silt or clay, which are very fine **[1]**.
2. **Any suitable answer which gives some idea of the action of abrasion and then a result, e.g.** Moving water will use the material / particles to scour / scrape the river bank below the surface. The base of the bank will be undercut and the top of the bank may become unstable and collapse. **[3] [Maximum 2 marks for only the process or result]**
3. **Any suitable answer with two points explained, e.g.** Drier time / drought **[1]** means there is less water in the channel so its ability to transport material is reduced **[1]**. The channel gets wider **[1]** so there is more friction and a reduction in available energy for transport **[1]**. Channel is less steep **[1]** so less momentum and so less energy available for transport **[1]**. **[Up to 4 marks]**

Page 87: Rivers 2: Landforms

1. **Any suitable answer that makes the link between available energy and water movement, e.g.** Small rivers do not have a great deal of energy available (most is used to move the water along) **[1]** so they take an efficient course around obstacles / wind around obstacles **[1]** because they cannot erode the land / erode sideways **[1]**.
2.

[1]

Page 87: Rivers 3: Flooding and Management

1. A diagram / chart showing discharge **[1]** in a river following a rainfall event **[1]**.
2. **Any suitable answer which shows an understanding of the water table and the related ability of an area to hold rainfall, e.g.** A high water table means that the ground will not be able to hold much more water **[1]**, so after further rainfall more water is likely to move towards the channel **[1]**. This may cause the channel to overflow / go overbank / flood **[1]**.
3. **Any suitable answer which states the activity / structures and shows how it was / they were part of the cause of a flood, e.g.** York: More impermeable surfaces **[1]** from increase in size of built-up area **[1]** means water gets into channels quicker **[1]**. Water from Yorkshire Dales rivers gets into the Ouse more quickly **[1]** because farming has changed / rainwater isn't being held on farmland as much **[1]**. Erosion of uplands by walkers **[1]** has removed peat which used to hold a lot of water **[1]**. **[Up to 3 marks]**
4. **Any suitable answer which states the place (region or town / city) and gives specific information about a particular storm / flood event, e.g.** York: It is low lying at 15 m **[1]** so water from surrounding highlands could gather there **[1]**. It has two rivers running through it, the Foss and Ouse **[1]** so is doubly vulnerable if discharges rise **[1]**. Upstream from York, the Ouse is joined by the Swale, Ure and Nidd **[1]** – three rivers from the Yorkshire Dales **[1]**, which is an area of high rainfall **[1]**. On December 4 and 5, 2015, the city had flooded, so land was saturated **[1]**. Further rain through the month led to flooding on December 26 **[1]**. **[Up to 4 marks]**

Page 88: Glaciation 1: Processes

1. On top of the ice **[1]**
2. A – En-glacial moraine **[1]**; B – Terminal moraine **[1]**; C – Sub-glacial moraine **[1]**
3. Deposition means that the glacier does not have enough energy to carry all the load **[1]**. There is less energy if the ice gets thinner **[1]** or if there is melting at the sides and front **[1]**, or if a glacier moves from high to low ground **[1]**.

Page 88: Glaciation 2: Landscape

1. On the floor of a U-shaped valley (trough). **[1]**
2. Glaciers form in tributary river valleys and flow into the main / spine valley **[1]**. More ice enables more erosion, so more tributaries mean that more ice is available **[1]**.
3. Abrasion on the back wall helps to steepen it **[1]**, then to deepen the floor **[1]**. As ice leaves the corrie, abrasion smooths and shapes the threshold **[1]**.

Page 88: Glaciation 3: Land Use and Issues

1. **Any named upland glaciated area with four features of the climate described, e.g.**
Snowdonia: wet with about 2600 mm rain per year; cool summer with average temperatures 14–15°C; winter average temperature is 3–4°C but nights and some days are below zero; minimum 60 days of frost; strong winds. **[4]**

Or Lake District: wet with 2000 mm rain; cool summer with average temperature 15°C; winter average is about 4°C but colder in the high mountains; over 60 days of frost; strong winds. **[4]**
Or Scottish Highlands: wet with 2000 mm rain / snow; very cool summer with average 12°C; winter average is about 1°C; 160 days of frost; good chance of snow; strong winds. **[4]**

2. **Any suitable answer with linked points, e.g.** Sheep can compact the soil **[1]**, especially if there is overgrazing **[1]**, so that rain does not infiltrate **[1]** and water carries sediment downhill **[1]** into rivers, which can cause flooding **[1]**. Fertilisers to improve soils **[1]** for crops or pasture can wash into lakes or rivers **[1]**, changing the pH **[1]** or affecting native plants and animals **[1]**. **[Up to 4 marks]**

Pages 89–91 Practice Questions

Page 89: Urbanisation

1. a) 10% **[1]**   b) 56% **[1]**   c) 68% **[1]**
2. The uncontrolled expansion of urban areas (towns and cities) **[1]**. It is considered a problem as it produces deserts of suburban housing **[1]**, reduces productive farmland and destroys wildlife habitats **[1]**.
3. a) **Any three from:** Unemployment; low wages; farming is difficult and unprofitable; few job opportunities; lack of social amenities; isolation; natural disasters. **[3]**
   b) **Any three from:** More job opportunities; higher wages; better schools and hospitals; better housing and services (like water, electricity and sewerage); better social life; better transport and communications. **[3]**
4. **Any suitable answer which focuses on four differences, e.g.** The rate of growth (i.e. fastest in LICs); rural-to-urban migration (reference to push-pull factors); industrialisation (as countries industrialised, people began to migrate to cities for jobs; as this happened in HICs over 100 years ago, they now have large urban populations; LICs are still in the early stages of industrialisation, and so urban populations are still growing); growth of shanty towns (on periphery of LIC cities, but not a feature of HIC cities); primate cities in LICs (LICs often have one primate city rather than a number of large cities of similar size). **[Credit can be given for use of correct geographical terminology, e.g. HIC, LIC, NEE, millionaire cities (growing rapidly in LICs), megacities (mainly found in LICs). Credit can be given for relevant examples.] [Up to 4 marks]**

Page 89: Urban Issues and Challenges 1

1. **Any suitable answers, e.g.** Better healthcare; better schools; better housing; jobs; wages; better social services **[4]**
2. **Any suitable answer, e.g.** Lack of clean water supply; overcrowding; lack of job opportunities; crime rates; lack of services **[4]**
3. **Any suitable answer, e.g.** Beaches; Sugar Loaf Mountain; statue of Christ the Redeemer; Rio Carnival; sports events **[4]**
4. **Any suitable answer which refers to two different social challenges [2], with further development of each point needed [2], e.g.** Migration (rapid growth of Rio in recent years from the movement of people into the city from surrounding rural areas putting pressure on services and amenities); housing (many rural migrants begin their life in the city in shanty towns called favelas; these are unplanned and spontaneous, often growing up on poor quality land; favelas are overcrowded, residents have no legal ownership, and houses are built of cheap materials; these are areas lacking clean water and sewage disposal, with no schools or hospitals, and few job opportunities); healthcare (favelas have high levels of disease and illness, with poor quality healthcare – particularly for maternity and care of the elderly); education (nearby schools suffer from poor funding, low enrolment due to poverty, and a lack of trained teaching staff); water supplies (clean water is not available for 12% of the population in Rio's favelas); energy (a shortage of available electricity supply means frequent blackouts for residents); crime (street crime is high, with powerful drugs gangs controlling the favelas).

Page 89: Urban Issues and Challenges 2

1. Docks converted into marinas **[1]** with housing and offices and small businesses. **[1]**
2. **Any suitable answers, e.g.** Housing; sports facilities; improved transport links; the Queen Elizabeth Park **[4]**
3. **Any suitable answers, e.g.** Wider use of diesel-electric hybrid buses **[1]** and the trialling of hydrogen fuel cell buses and combined diesel and biofuel buses **[1]**.

4. **Any suitable answers, e.g.** The Docklands Light Railway (DLR) **[1]** connected the area to the London Underground system **[1]**; London City Airport **[1]** specialises in STOL (short take-off and landing) air travel **[1]**; high speed rail link (HS2) from London to Birmingham (and eventually Manchester) **[1]**; Crossrail project from Paddington Station to Reading in the west and Abbey Wood in the east of the city **[1]**; plans for a new runway at Heathrow **[1]**. **[Up to 4 marks]**

Page 90: Urban Issues and Challenges 3

1. 90% **[1]**
2. **Any suitable answers, e.g.** Food and water; fuels and energy; building materials and consumer goods. **[2]**
3. **Any suitable answer which refers to three different strategies [3], with further development (but not necessarily explanation) of each point needed [3], e.g.** Car sharing (resulting in fewer vehicles on the road; it is particularly effective when work colleagues share, as rush hour traffic is reduced); vehicle-restricted areas (that reduce the number of vehicles on roads and reduces vehicle movements in busy, congested areas); increased use of public transport (including buses, trams and trains, organised into an integrated transport system; this helps to reduce the number of vehicles on the roads); bus lanes (improve traffic flow); park-and-ride systems (reduce vehicles on roads in busy, congested areas, and improve traffic flow by reducing need for parking); increased bicycle use (reduces the number of vehicles on roads); new road designs and traffic controls (such as one-way systems and traffic controls to improve traffic flow). **[Up to 6 marks]**
4. **Any suitable answer which refers to a range of different strategies [up to six for 6 marks], with further development of individual points needed to gain additional marks (up to a maximum of 6 in total). A suitable answer could include a variety of points that help a city to become carbon neutral (produces as much energy as it consumes):** Renewable energy for housing (e.g. solar panels – to generate electricity and reduce carbon emissions); reduce energy for industry (e.g. solar panels, wind generators, hydro-electricity – to generate electricity and reduce carbon emissions); reduce energy use at home and in the workplace (through the use of renewable energy sources, efficient insulation, double and triple-glazing, water and electricity meters, low-flow taps, etc.); reduce household and industrial waste (through increased recycling of waste materials, e.g. cardboard, metals, glass, food waste); energy from waste (composting of food and green waste helps reduce methane emissions); use of greywater systems.

Page 90: Measuring Development and Quality of Life

1. **Any suitable answers, e.g.** Birth rate; life expectancy; literacy; infant mortality; access to healthcare **[2]**
2. **Any suitable answers, e.g.** Gross domestic product (GDP); wages; trade figures **[2]**
3. **Any four from:** Stable growth rate; low birth rates; low death rates; ageing population; some HICs may even have declining populations **[4]**

Page 90: The Development Gap

1. **Any two from:** Extreme climates make food production difficult in some areas; land in some regions is unsuitable for farming (e.g. too high, too steep, thin soils); limited water supplies in some regions; natural hazards (e.g. tropical storms and earthquakes) affect some regions more than others; an imbalance in the location of natural resources (e.g. precious metals, energy sources). **[4]**
2. **Any suitable answers, e.g.** A fair price for goods produced (e.g. cocoa) means more money to spend on food, schooling and improving family life **[2]**; co-operatives provide extra services (e.g. healthcare) **[2]**
3. **Any suitable answers, e.g.** Wildlife safaris (the 'big five'); tribal culture; warm climate with all-year sunshine; varied landscapes; spectacular national parks **[4]**
4. **Any suitable answers, e.g.** Tourism provides 15% GDP; creation of tourism-related jobs; efforts made to preserve tribal cultures; better protection of landscapes; greater development of infrastructure **[4]**

Page 91: Changing Economic World Case Study – Vietnam

1. **Any suitable answers, e.g.** Cheap labour; longer working hours; fewer health and safety laws; less stringent environmental laws **[3]**
2. Profits from a transnational corporation are sent back to the home country. **[1]**
3. **Any suitable answer, e.g. for Vietnam:** Approximately 37% of people work in the primary sector **[1]**, in jobs such as farming, mining and forestry **[1]**; around 24% of people work in the

secondary sector in manufacturing [1], such as for companies like Nike making clothes and footwear [1]; around 35% of people work in the tertiary sector in services [1], such as in education, health, shopping [1]; tourism is a rapidly growing tertiary industry in Vietnam [1]. **[Up to 3 marks]**

Page 91: UK Economic Change

1. Any suitable answers, e.g. Bristol [1]; Cambridge [1]
2.

2001	2020 – higher, same or lower?	
Primary = 1%	Same	[1]
Secondary = 24%	Lower	[1]
Tertiary and Quaternary = 75%	Higher	[1]

3. Any suitable answer, e.g. Improvements to transport like HS2 [1], which will make the north a more attractive place to live [1]; it will also attract business investment [1], creating jobs [1]. The Government has set up enterprise zones [1] to attract business to locate in the north [1], creating jobs and economic growth [1]. The Government is giving money and decision-making powers to northern cities [1], ensuring that decision-making is suitable for the economy and people in the north [1]. **[Up to 3 marks]**

Page 91: UK Economic Development

1. Any suitable answers, e.g. Germany [1]; USA [1]; China [1]; EU [1]
2. Any suitable answer, e.g. Using renewable sources of energy like wind and solar [1] to reduce carbon dioxide emissions [1] (which contribute to the enhanced greenhouse effect and climate change [1]). Reducing packaging [1] means there is less waste to be processed [1]. Designing buildings that blend in with the environment [1], e.g. Adnams in Suffolk has a green roof and lime / hemp walls [1].
3. A greenfield site is a plot of land that has not been used before [1], whereas a brownfield site is land that has been previously used for development [1].

Pages 92–111 Revise Questions

Page 93 Quick Test

1. The UK is unable to grow enough food for its population and there is a great demand for food out of season.
2. Because of a growing population and lifestyles that use more water.
3. It is the driest part of the country but also the area with the highest population.
4. Declining fossil fuel reserves (e.g. gas) mean that the UK has to import more, while more affluent lifestyles mean that households use more energy.

Page 95 Quick Test

1. A country having reliable access to a sufficient quantity of affordable, nutritious food to feed its people.
2. Any suitable answers, e.g. Aridity; floods; extreme cold / heat; water availability; soil type
3. Any suitable answers, e.g. Civil wars; poor techniques leading to degradation and desertification; poor distribution networks; lack of adequate storage.

Page 97 Quick Test

1. Any suitable answer, e.g. The use of new farming methods, such as hydroponics and genetically-modified crops, that supply larger yields; in LICs, examples include high-yielding varieties of cereals and modern farming techniques such as irrigation, synthetic fertilisers and pesticides.
2. Any suitable answer, e.g. Eating sources of food that are locally grown and only eating fruit and vegetables that are in season; changing diets to those less reliant on meat and dairy products; using sustainable fish sources and reducing food waste in homes and shops.
3. Ethical consumerism is the purchase and use of sustainable goods, such as food that is locally grown and only eating fruit and vegetables that are in season.

Page 99 Quick Test

1. The ability of a population to ensure access to quantities of safe water to maintain life, social well-being and economic development.
2. Any suitable answer, e.g. Greater demand from agriculture, industry and domestic properties; climate change; over-extraction by man.
3. Poor sanitation leads to the pollution of water supplies (this problem increases with a rise in population).

Page 101 Quick Test

1. Methods of conservation can be used in the storage of water; expensive schemes, such as water transfer schemes and desalination plants, can be used.

2. Any suitable answers, e.g. Using low-flush toilets; taking short showers; only using washing machines and dishwashers when full.
3. Any suitable advantages, e.g. The quality of the water is very good and needs less treatment than river water to make it safe to drink; it stays available during the summer and during droughts when rivers and streams have dried up.
Any suitable disadvantages, e.g. The use is not sustainable as the groundwater takes a long time to replenish.

Page 103 Quick Test

1. Eastern Europe, including Russia – have large reserves of natural gas and coal; Middle East and North Africa – have large oil reserves.
2. Energy usage is high in countries that have either hot or cold climates due to heat or air-conditioning use; usage is also high in countries with a great annual range of temperature.
3. Agriculture uses oil to power machinery, for transport and in agricultural chemicals, so an increase in oil price causes a rise in food prices.

Page 105 Quick Test

1. Gas produces only half as much carbon dioxide as coal and, with the introduction of new exploitation methods like fracking, is relatively cheap.
2. Wind power is the most widely used renewable in the UK because it experiences windy conditions for much of the year. More than 11 000 onshore and offshore wind turbines produce about one quarter of the UK's electricity output.

Page 107 Quick Test

1. The greenhouse gas emissions caused by an organisation, event, product or individual.
2. Any suitable answers, e.g. The use of LED lighting; motion sensors on lighting to turn it off when no-one is around; double glazing and large south-facing windows to let in more light.
3. Any suitable answers, e.g. The introduction of park-and-ride schemes; bus and cycle lanes; congestion charging; trams; hybrid buses; bikes for hire.
4. Any suitable answers, e.g. The use of hybrid and more efficient engines in vehicles; electric vehicles; aircraft that are more fuel-efficient; vehicles that are more aerodynamic.

Page 109 Quick Test

1. A hypothesis is an idea or explanation for something that has not yet been proved.
2. Primary data is information collected by you and your fellow students. Secondary data is information you have found on a website or in a book (i.e. which someone has collected previously).
3. Quantitative data includes statistics and numbers (i.e. things that can be counted or calculated). Qualitative data is non-numerical and can be subjective (i.e. things such as field sketches, opinions, news articles).

Page 111 Quick Test

1. Synoptic thinking means using the knowledge, understanding and skills that you have learned throughout the course to help answer questions about a new situation that you haven't studied before. Critical thinking is interpreting, analysing and evaluating ideas and arguments based on the information that you are given and your existing knowledge and skills.
2. a) Stakeholders are people or groups with a real interest in an issue or in the development of a decision or policy, e.g. the residents of an area where a new road is to be built.
   b) Key players are people or groups with an interest and a significant degree of control about an issue or in the development of a decision or policy, e.g. the company paid to build the road, or the politician who gives it the go-ahead.
   c) Interest groups are organisations of people with a common cause about which they attempt to influence policies or decisions, without seeking political control, e.g. if the residents of the area where the new road is to be built formed a group, it would be an interest group.
3. 'Top-down' management is where the person / organisation in charge makes all the decisions and tells those lower down exactly what to do. A 'bottom-up' approach takes into account everyone's opinion and the decision is formed by the whole group.
4. Measures that provide people with a degree of control over problem-solving or decisions.

Pages 112–114 Review Questions

Page 112: Urbanisation

1. Any simple definition of urbanisation, making reference to an increase in the proportion of people living in urban areas (or towns / cities / built-up areas). [1]

2. Any suitable answer with two different reasons. A suitable answer should refer to any 'pull' factor of rural-to-urban migration, such as industrialisation creating a demand for jobs, better healthcare, better schools, etc. or any 'push' factor of rural-to-urban migration, e.g. unemployment, difficulties in farming, low wages, etc. [2]
3. a) False [1] b) True [1] c) False [1] d) True [1] e) False [1]
4. a) Push factor [1] b) Pull factor [1] c) Push factor [1]
   d) Push factor [1] e) Pull factor [1]

Page 112: Urban Issues and Challenges 1

1. **A suitable answer could include:** Provision of housing for rural migrants (instead of allowing uncontrolled growth of favela housing); upgrading of favelas (by providing pavements, electricity and sewage systems through schemes like the Favela Bairro Project); promotion of self-help building schemes (where local residents are provided with building materials); giving residents legal rights of ownership or low rents on properties (to develop permanent communities); improved transport systems (to give residents better access to work in the city); encouraging creation of businesses like shops and restaurants (to provide employment and allow favelas to be self-contained); improved law and order through pacification programmes (to reduce crime in the favelas); encouraging tourism and creation of tourist-related businesses (to bring additional income opportunities to residents). **[Two different ways that quality of life can be improved need to be identified [2], with further development required [2]]**

2. **A suitable answer could include:** Strength of community; creation of new businesses (e.g. shops, restaurants, tourist-related businesses, cottage industries like pottery); residents sending money home to families in the countryside; recycling (with particular reference to use of building materials and waste recycling); cultural and ethnic diversity; growth in tourist-related activities. **[Three different positive aspects need to be given [3]]**

3. **A suitable answer could include:** Urban sprawl (as the city continues to grow, it encroaches on surrounding rural areas); pollution (air pollution from heavy traffic); traffic congestion (in the city centre); sea pollution (from sewage and industrial waste); waste disposal (particularly in the favelas, many of which are inaccessible to collection vehicles). **[Identification of one environmental challenge [1], with further development required [1]]**

4. **A suitable answer could include:** Favelas being unplanned and spontaneous (often growing up on poor quality land); overcrowding (with many houses crammed into a small area); poor quality housing (often built from cheap materials like wood or corrugated iron); lack of services (like clean water, sewage disposal, electricity supply); poor health (due to lack of clean water and lack of sewage disposal); few job opportunities (leading to poverty); few facilities (such as schools, hospitals and public transport); high levels of crime (particularly street crime, with powerful drugs gangs controlling many areas). **[Reference needed to a range of different living conditions for up to 4 marks), with further development of individual points needed to gain additional marks (up to a maximum of 4 in total).]**

Page 112: Urban Issues and Challenges 2

1. **A suitable answer could include:** Dereliction (as manufacturing industries have declined, particularly in the inner city, much land has been left in a state of dereliction); waste disposal (a large city produces a lot of household and commercial waste for disposal); pollution (atmospheric pollution from industry and vehicles). **[Two different environmental challenges identified [2], with further development of individual points [2]]**

2. **A suitable answer could include:** Crossrail from Reading to Abbey Wood and Shenfield; third terminal proposal for Heathrow Airport; diesel-electric hybrid buses; congestion zone to reduce pollution [4]

3. **Any suitable answer, e.g.** Historic buildings; monuments; sporting events; cultural events; entertainment (credit can be given for named relevant examples). **[Up to 3 marks]**

4. **A suitable answer which includes three social challenges, e.g.** Housing inequalities (provision of more affordable housing); need to improve education standards to correct variation in performance between boroughs; poor access to healthcare for people with poor English language skills; increased in-migration (particularly international migration) attracted by finance and knowledge-based employment opportunities. **[Up to 3 marks]**

Page 113: Urban Issues and Challenges 3

1. **Any two from:** Sewage; exhaust gases; household waste; industrial waste; building waste. [2]

2. **Any suitable answers, e.g.** The use of hydrogen-fuelled [1] and electric cars [1] helps to reduce air pollution and cuts down on reliance on fossil fuels.

3. **Any suitable answer with development of each point made, e.g.** Renewable energy use such as solar panels (which reduces the requirements for energy from fossil fuels and reduces carbon emissions) [1]; insulation (of walls, lofts, etc. which reduces heat loss and therefore reduces energy use) [1]; affordable prices and rents (which supports a balanced community of different income levels, and reduces demand on local authority housing) [1]; passive energy use (to reduce demands on energy produced from fossil fuels and reduce carbon emissions) [1].

4. **A suitable answer could include:** Urban greening (a target of 40% green space of parklands and trees help to meet leisure needs of residents and absorb carbon dioxide from the atmosphere); urban forests (e.g. Adelaide in Australia – they help to meet leisure needs of residents and absorb carbon dioxide from the atmosphere); urban architecture (such as vertical farming, allotments and roof-top gardens help to increase food production and reduce 'food miles' from imported products); use of brownfield sites for development (to avoid the need to expand into untouched greenfield sites at the edge of the urban area); greenbelt (to maintain an area of green space in easy reach of all residents within the urban area). **[Two different environmental improvements need to be identified [2] with further development of individual points needed [2]]**

Page 113: Measuring Development and Quality of Life

1. HDI combines life expectancy [1], literacy and income to measure development [1]. It has become popular as it does not rely on one single indicator (usually wealth). [1]

2. **Any suitable answer with two clear reasons to show how a NEE differs from a LIC, e.g.** Country wealth (a LIC is classified by the World Bank as one with less than $1085 GNI per capita); rate of economic development (NEEs are beginning to experience high rates of economic development, usually with rapid industrialisation); reliance on agriculture (NEEs no longer rely primarily on agriculture in their economy. (Credit can be given for naming relevant examples.) **[Up to 2 marks]**

3. **Any three from:** Population increasing rapidly; high birth rates; falling death rates; many young dependants [3]

Page 113: The Development Gap

1. **Any suitable answers, e.g.** Corrupt governments [1]; war and conflicts [1]

2. **Any suitable answers which include:** A description of intermediate technology (a form of aid, sometimes also known as 'simple' or 'appropriate' technology that does not necessarily use the most up-to-date technology available to improve a situation in a LIC) [2]; and for identifying how intermediate technology can reduce the development gap (the difference in standard of living between rich and poor countries) through making improvements to living or working conditions in a LIC [2]. (Credit can be given for reference to examples of intermediate technology aid projects.)

3. Debt reduction occurs when countries are released from some of their debts [1], although often with other conditions attached [1]. Debt relief is when lending countries write off debts [1] that have spiralled out of control due to high interest payments. [1]

4. **A suitable answer could include:** Money generated goes to large overseas companies (rather than benefiting local people); tribes forced off their land to create national parks (causing changes to lifestyles and loss of cultural integrity); dress and behaviour of tourists can offend local people (particularly certain local religions); safari vehicles increase in number and drive off-road to locate wildlife (disturbing wildlife and damaging natural vegetation); coral reefs damaged by divers and tourist boats (causing destruction to reefs which then fail to regenerate); threat of terrorism **[4 marks for identifying four different disadvantages, or fewer disadvantages can be given with additional marks added for development of individual points]**

Page 114: Changing Economic World Case Study – Vietnam

1. A suitable answer could include:

   Advantages: secondary sector employment opportunities (particularly for women, providing new skills and higher wages); cheap labour supply (for the TNC); growth in the home market (adding to exports and making a positive contribution to GDP); fewer industrial laws and restrictions (makes operations for the TNC easier); extra tax revenue (that can be spent by the host country on infrastructure improvements); lead to attraction of other TNCs ('snowball' effect brings strength to the host country's economy). **[Up to 3 marks]**

   Disadvantages: company image and advertising (can undermine culture of host country); political influence held over host country government (can lead to undemocratic decisions); investment easily moved away from host country (can leave unemployment and a vacuum in the host country economy); environmental impact (pollution from factories). **[Up to 3 marks]** **[Advantages and disadvantages could apply to either a TNC itself, or the host country. Marks can be awarded for each individual factor identified, and additional marks can be gained for further development of points made.]**

2. Improving transport infrastructure enables people to access services like education and healthcare more easily **[1]**. It also encourages businesses to develop **[1]** and can attract foreign transnational investment **[1]**. Training workers provides people with new skills **[1]**, enabling them to get work **[1]** and improve their quality of life **[1]**. Promoting women's economic empowerment encourages women into the workforce **[1]**, enabling them to earn money and improve their quality of life **[1]**. It also raises the status of women **[1]**. Tackling hunger means people are healthier **[1]**. This reduces demand for healthcare **[1]** and people can work and earn money **[1]**. **[Up to 3 marks]**

3. Industrial development can cause air pollution **[1]** and can affect people's health, e.g. asthma **[1]**. Waste and toxic chemicals cause land and water pollution **[1]**, which can affect public drinking water supplies **[1]**. Increasing the use of fossil fuels releases greenhouse gases that cause the enhanced greenhouse effect **[1]**. This causes climate change and global warming **[1]**. Deforestation caused by farming and logging has resulted in the loss of animal habitats **[1]**. The soil is left exposed, resulting in erosion **[1]**, and the land becomes less productive for farming **[1]**, meaning crop yields reduce **[1]**. Economic growth has seen people's incomes rise **[1]**. There are now less people living in poverty **[1]**, meaning they can access healthcare and education **[1]**. They can also afford better housing **[1]**. **[Up to 5 marks]**

Page 114: UK Economic Change

1. a) False **[1]** b) True **[1]** c) True **[1]** d) False **[1]** e) False **[1]**
2. The number of people employed in farming and mining has declined due to mechanisation **[1]**, so the jobs that people used to do are now done by machines, e.g. combine harvesters in farming **[1]**. Numbers employed in mining have also decreased due to raw material, such as coal, running out **[1]**. Secondary industry has declined due to competition from overseas **[1]**. Cheap labour costs in countries such as Vietnam mean that goods can be made more cheaply than in the UK **[1]**. Some of the production methods used in the UK are outdated **[1]**, meaning UK industries are not competitive compared to those overseas **[1]**. **[Up to 3 marks]**
3. **Any suitable answer, e.g. Bramhall, south of Manchester:** Social changes – population has increased **[1]**; good community spirit with a range of events throughout the year **[1]**; old housing has been modernised **[1]**; local schools are oversubscribed **[1]**; many services are aimed at wealthier people **[1]**; journey times are slow due to congestion **[1]**. Economic changes – new businesses have set up **[1]**; new services have opened due to the growing population **[1]**; jobs have been created in local businesses and services **[1]**; house prices have increased, forcing some people out of the area **[1]**. **[Up to 5 marks]**

Page 114: UK Economic Development

1. UK, USA, Canada, France, Germany, Italy and Japan **[3 if all correct; 2 for up to five correct; 1 for up to three correct]**
2. **Any suitable answers, e.g.** Road building and improvement programmes (including 'smart' motorways) **[1]**; rail network upgrades (including HS2 link) **[1]**; airport expansions (e.g. Heathrow) **[1]**
3. **Any suitable answer, e.g.** The UK trades with lots of countries such as China and the USA **[1]**. Importing and exporting goods and services connects the UK with other countries **[1]**. The UK is a member of the G7 **[1]**, which means it works with other countries to make global decisions **[1]**. People in the UK have good access to the internet **[1]**, which enables them to communicate with friends and family in other countries **[1]**. It also means people can buy goods and services from other countries **[1]**. The UK has many international transport hubs, such as airports and the Channel Tunnel **[1]**, which enables people to travel abroad for business or tourism **[1]**. **[Up to 5 marks]**

Page 115: Overview of Resources – UK

1. The distance that food is transported from where it is produced **[1]** until it reaches the consumer **[1]**.
2. With increasing affluence, water demand is increasing as new housing is built with more than one bathroom **[1]**, and consumers demand labour-saving devices such as dishwashers and washing machines **[1]**. Industry and agriculture are also large users of water **[1]**.
3. **Any suitable answer, e.g.** Water is transferred from wetter areas to those drier areas with greatest demand **[1]**, using pipelines, aqueducts and rivers **[1]**. Other methods of conservation can be used in the home, such as using showers, low-flush toilets and 'green' appliances **[1]** or water can be recycled using greywater harvesting **[1]**. **[Up to 3 marks]**
4. **Any suitable answer with at least two advantages and two disadvantages, e.g.**

   Advantages: Fracking allows drilling firms to access difficult-to-reach resources of oil and gas **[1]**. It may help to reduce gas prices **[1]**. Electricity can be generated at half the carbon dioxide emissions of coal **[1]**.

   Disadvantages: The use of fracking has prompted environmental concerns **[1]**. It uses huge amounts of water that must be transported to the fracking site at significant environmental cost **[1]**. Cancer-causing chemicals are used and may escape and contaminate groundwater around the fracking site **[1]**. There are worries that the fracking process can cause small earth tremors **[1]**. Environmental campaigners say that fracking is simply distracting energy firms and governments from investing in renewable sources of energy **[1]**. **[Up to 5 marks]**

Page 115: Food 1

1. A lack of food for the population of an area, causing illness and / or death through starvation. **[1]**
2. **Any suitable answers, e.g.** Increase the number of irrigation schemes; restructure farming from subsistence to commercial farming; introduce new high-yielding varieties of staple crops; introduce better storage and transport systems **[3]**
3. **Any suitable answer with two physical and two human factors, e.g.** Physical factors: soils – fertility influences the type of farming; relief – flat land is preferable for arable farming; climate is the most important physical factor **[2]** Human factors: cost of land; market inertia – farmers may have farmed in a certain way and may be reluctant to make changes; governments (grants, tax barriers and subsidies); technology, e.g. genetically-modified crops **[2]**
4. **Any suitable answers, e.g.** Climate change is likely to hit the countries of Africa the worst **[1]**. Temperatures are likely to rise by more than 2°C **[1]** and this will lead to periods of heatwaves and drought **[1]**, causing desertification **[1]** in some areas and heavy rainfall causing flooding in others **[1]**. This could result in less land available for food production **[1]**. This may lead to increased rates of migration to cities **[1]**. **[Up to 4 marks]**

Page 115: Food 2

1. The application of water to increase the yields of crops, especially in those areas where water is in short supply. **[1]**
2. The 'Blue Revolution' is the growth of aquaculture **[1]** as a very highly productive way of producing food such as aquatic animals and plants in both salt and fresh water **[1]**. Fish farming, such as shrimp or salmon farming **[1]**, and the gathering of seaweed are all methods of aquaculture **[1]**. **[Up to 2 marks]**
3. **Any suitable answers, e.g.** Hydroponics is a method of growing plants using nutrient-rich water **[1]**, often using inputs of ultra-violet light **[1]**, such as high-value crops like tomatoes in glass-houses **[1]**. **[Up to 2 marks]**

Aeroponics is a similar way of growing plants in an air or mist environment [1] without the use of soil [1]. It has been used by NASA as a possible way of feeding astronauts on long space journeys [1]. **[Up to 2 marks]**

Biotechnology uses biological processes to develop new products [1] that can help feed the hungry by producing higher crop yields [1]. It can lower the input of agricultural chemicals into crops [1] with less vitamin and nutrient deficiencies [1] and free of allergens and toxins [1]. **[Up to 2 marks]**

Appropriate (or intermediate) technology is technology suited to the social and economic conditions of the people using it [1]. It is environmentally sound [1] and promotes sustainability as it uses locally sourced materials [1]. Examples include digging boreholes to supply water to irrigation schemes [1] or collecting animal dung to convert into methane for gas stoves to use instead of firewood [1]. **[Up to 2 marks]**

4. **Any suitable answers, e.g.** The Gezira Scheme in Sudan [1] is an example of a large-scale irrigation scheme where the flood waters of the Nile [1] are used to grow crops like cotton and wheat [1], but large amounts of water are taken out that affect other irrigation schemes downstream [1]. The cotton is a valuable export for the Sudanese economy [1] but attempts to diversify the type of crops have failed as the country has been affected by civil war [1]. **[Up to 5 marks]**

Page 116: Water 1

1. The inability of a country to ensure sufficient access to quantities of safe water [1] to maintain life, social well-being and economic development [1].

2. **Any two from:** The supply of water is uneven as the distribution of rainfall varies from place to place; arid places have a deficit of water; rising populations increase the demand for water for drinking, bathing, agriculture and industry. **[2]**

3. **Any suitable answer, e.g.** With increasing economic development, the demand for water also rises [1] as domestic use increases through more labour-saving devices like washing machines and dishwashers [1]. Industrial development takes place, leading to a greater demand for water in manufacturing [1] and in electricity generation [1]. Agricultural use of water also increases due to a greater demand for meat and dairy products [1]. **[Up to 4 marks]**

Page 116: Water 2

1. Schemes that move water from one river basin where it is available, to another basin where water is less available. **[1]**

2. Removing salt from either seawater or groundwater to make it drinkable. **[1]**

3. **Any suitable answer, e.g.** The South–North Water Transfer Project in China [1] will enable 44.8 billion cubic metres of water per year [1] to move from the Yangtze River in southern China [1] to the Yellow River Basin in arid northern China [1] through a system of canals and aqueducts [1]. The cost of the project is over $60 billion [1] but is likely to cause water shortages in some parts of China and pollution on some rivers [1]. **[Up to 5 marks]**

4. **Any suitable answer, e.g.** Groundwater makes up nearly 30% of all the world's fresh water [1] and is often used in places where there is not enough water to drink [1]. Groundwater quality is usually very good [1] and needs less treatment than river water to make it safe to drink [1], as the rocks through which the groundwater flows help to remove pollution [1]. Groundwater also responds slowly to changes in rainfall [1], and so it stays available during the summer and during droughts when rivers and streams have dried up [1]. It does not require large costs or technology (as it is often drawn from wells) or expensive reservoirs to store water in before it is used [1]. **[Up to 5 marks]**

Page 116: Energy 1

1. Between 1970 and 2020, the use of oil dropped from 48% to an insignificant amount [1], while the use of gas increased from 8% to 36% [1] and coal dropped significantly from 38% to 2% [1].

2. a) 43% **[1]**
   b) Improvement in technology has made the harnessing of these sources possible [1]; investment and subsidies available for installation of wind turbines and solar panels [1]; less reliance on fossil fuels needed to increase energy security [1].

3. North America has large coal resources but its large conventional oil resources are largely exploited [1]; it is now exploiting non-conventional oil and gas reserves but its huge energy consumption often outweighs supplies [1]. Europe is heavily dependent on energy imports [1] as it has declining fossil fuel supply [1].

Page 117: Energy 2

1. **Any suitable answer with examples, e.g.** The capture of energy from existing flows of energy [1], such as sunshine, wind, wave power, flowing water (HEP), biomass and geothermal heat [1].

2. Coal is the dirtiest of fuels (producing 31% of all carbon dioxide) so technology has been developed that removes much of this carbon dioxide before the coal is burned [1]. The reason for this is the impact of greenhouse gases like carbon dioxide on climate change [1]. Laws have been passed to cut down the emissions of coal-fired power stations [1].

3. **Any suitable answer, e.g.** The Muppandal wind farm in Tamil Nadu in southern India [1] has over 3500 wind turbines and is one of the largest in the world [1]. It produces enough electricity to power a million homes [1]. The turbines were purchased by wealthy people, who bought these as an alternative to paying high taxes [1]. As India is a heavy user of coal, installations such as Muppandal reduce the amount of carbon dioxide and the pollution produced [1].

4. **Any suitable answer with four advantages and four disadvantages, e.g.** **[8]**

Advantages	Disadvantages
Geographical limitations: nuclear power plants don't require a lot of space. [1] Nuclear power stations do not contribute to carbon emissions or pollution. [1] Nuclear energy is by far the most concentrated form of energy; a lot of energy is produced from a small amount of fuel. [1] Nuclear power is reliable; it does not depend on the weather. [1] We can control the output from a nuclear power station to fit our needs. [1] Nuclear power produces a small volume of waste. [1]	They have to be located near water for cooling. [1] Disposal of nuclear waste is very expensive: it is radioactive so it has to be disposed of in such a way as it will not pollute the environment. [1] Decommissioning (close down and dismantling) of nuclear power stations is expensive and takes a long time. [1] Nuclear accidents can spread 'radiation-producing particles' over a wide area – this radiation harms the cells of the body which can make humans sick or even cause death. [1] They could be targets for terrorism. [1]

Page 117: Energy 3

1. **Any suitable answer, e.g.** Carbon footprints in HICs tend to be high because of the lifestyles people lead [1], with a reliance on fossil fuels [1], using electrical devices [1] and a diet that relies on food that is imported [1] or uses high levels of inputs derived from fossil fuels [1]. **[Up to 4 marks]**

2. **Any suitable answer, e.g.** In LICs, carbon footprints tend to be low as lifestyles are more sustainable [1] as less fossil fuel is used [1] and food tends to be locally produced [1]. **[Up to 2 marks]**

3. **Any suitable answer, e.g.** Energy can be conserved through the design of homes by the use of insulation in loft spaces and walls [1], double-glazed windows [1] and larger windows in south-facing walls [1]. **[Up to 2 marks]**

4. **Any suitable answer, e.g.** Using energy-efficient devices like kettles that only boil one cup of water [1], along with turning off appliances when not in use (e.g. not using standby buttons), help to conserve energy in the home, [1]. Other methods include the use of LED light bulbs [1], washing machines that wash clothes at 30°C or lower [1] and thermostats on central heating systems that can be controlled using a mobile phone [1]. **[Up to 4 marks]**

Page 117: Fieldwork

For all questions, you need to know the titles and be able to describe the location for both of your enquiries. The answers given below are suggestions for things you may consider depending on the nature of your enquiries.

1. a) and b)
   - Secondary data: cost-benefit data; comparing old and contemporary maps, aerial images, photographs; virtual fieldwork (e.g. Street View; geography.org.uk).
   - Primary data: environmental quality survey (to produce environmental quality index); land use mapping; questionnaires; oral histories; field sketches; photographs.

a) Specific to physical environment
   Rivers / Coast:
   - Secondary data: Government information such as flood-risk maps from Environment Agency; cost-benefit data; comparing old and contemporary maps, aerial images, photographs.
   - Primary data collection methods for depth, width, wetted perimeter, velocity, gradient, sediment size and shape (roughness) using equipment such as a tape measure, range poles, chain, metre ruler, clinometer, pantometer, hydro-prop with impeller, flow meter, floating object (e.g. table tennis ball, cork, orange, stick), stopwatch, callipers. Physical maps (e.g. geology, drainage, floodplain zones, coastal management zones). As appropriate, profiles or transects might be produced. **[Up to 1 mark]**

b) Specific to human environment
   Urban / Rural:
   - Secondary data: Government information such as census data from the Office of National Statistics; census data for age profile, unemployment, qualifications, housing tenure, house price or rental survey, Index of Multiple Deprivation (composite indicator), rural deprivation. Human geography maps (e.g. infrastructure maps; Goad maps; neighbourhood zones; planning maps).
   - Primary data collection methods for traffic counts; car age or origin; pedestrian counts; shopping quality (e.g. independents vs. chains); rate of turnover for retail premises; footfall; index of decay; building heights; noise levels or air quality (could use a smartphone app). **[Up to 1 mark]**

2. For both environments:
   - Ease of access physically and legally (e.g. land ownership); distance from school and transport availability; size of area or ward in which representative data can be gathered in time available.
   - Safety: risk assessment and measures to reduce risk; links to hypotheses about the area; ability to make comparisons between contrasting places or to observe changes since earlier investigations took place.
   Specific to physical environment
   - Rivers / Coast: Manageable size, e.g. small area in which students can walk up and down and, if necessary, be near water or steep slopes or cliffs easily and safely. **[Up to 1 mark]**
   Specific to human environment
   - Urban / Rural: Manageable size, e.g. small area(s) of a settlement that can be accessed easily and safely on foot; traffic. **[Up to 1 mark]**

3. Briefly explain the data collection method. 'Justify' means that you need to explain why it was effective and how it promoted your investigation. State whether the data was quantitative or qualitative; explain the advantages of such data. Describe and explain sampling techniques you used to ensure the reliability of your data, e.g. random, systematic, stratified. If you used GIS, explain why such georeferenced data is a useful technique. State how the data contributed to answering a key question or testing a hypothesis. **[Up to 3 marks]**

4. Briefly explain the limitations of a data collection method (e.g. why errors may have occurred) and how it might be improved. A good answer should consider questions such as these:
   If the sampling technique was a limiting factor, what alternative sampling technique might be used? If the sample size was unrepresentative, could more samples be taken? If the reliability of data is an issue due to when it was gathered, would the findings be more reliable if the investigation was repeated at a different time of day / different day of the week / different time of the year? If the reliability of data is an issue due to inconsistent data gathering (e.g. students not using equipment or techniques in the same way), would more training or practice be helpful? **[Up to 3 marks]**

Pages 118–120 Review Questions

Page 118: Overview of Resources – UK

1. A carbon footprint measures the total greenhouse gas emissions caused directly and indirectly **[1]** and is measured in tonnes of carbon dioxide **[1]**.
2. As the UK relies on fossil fuels and domestic supplies of coal, gas and oil are running out **[1]**, it means it has to import more from other countries **[1]**.
3. **Any suitable answer, e.g.** There are likely to be conflicts between the gas companies and local people in the areas where the gas is drilled **[1]**. This may be due to the impact of traffic **[1]**, pollution of water supplies **[1]** or the possibility of earth tremors **[1]**. There may also be conflict between gas companies and environmentalists **[1]**. **[Up to 3 marks]**
4. **Any suitable answer, e.g.** In the 1960s, most of the UK's electricity was generated using coal, oil and a small amount of nuclear **[1]**. Gas was also made from coal **[1]**. Now around 40% of the UK's electricity is made from fossil fuels (mainly gas) **[1]**, 40% from renewables and the rest from nuclear **[1]**. Coal-based energy has reduced dramatically over the last 50 years **[1]**. **[Up to 4 marks]**

Page 118: Food 1

1. When all people at all times have access to sufficient, safe, affordable nutritious food **[1]** to maintain a healthy and active life **[1]**.
2. **Any two from:** Climatic factors (too hot, too cold, too dry, too wet); water availability (for irrigation); soil type / fertility. **[2]**
3. **Any two from:** Population size (the ability of the country to support its people); insufficient farming skills; insufficient financial investment into a country's agricultural industry. **[2]**
4. **Any four from:** Farmers are poor and are stuck in the cycle of poverty; lack of investment in agriculture due to poverty; climatic factors including climate change and drought; civil war; price fluctuations leading to unstable markets; food wastage; poor storage and distribution networks **[4]**

Page 118: Food 2

1. **Any suitable answer, e.g.** Organic farming is a sustainable method of food production **[1]** that does not use synthetic pesticides or fertilisers **[1]**, instead using organic products to fertilise **[1]** (plant and animal waste) and biological solutions for pest and disease control **[1]**. **[Up to 2 marks]**
2. **Any suitable answer, e.g.** Reducing food waste and losses in both our homes and in shops could save UK consumers up to £2.4 billion a year. The average UK family throws away enough food for around six meals a week **[1]**. Buying less food in the supermarket and using goods before they go 'off' would lead to a more sustainable lifestyle **[1]**.
3. **Any suitable answer, e.g.** The Green Revolution is an initiative that has increased agricultural production **[1]**, particularly in LICs since the late 1960s **[1]**. The Green Revolution saved around a billion people from starvation **[1]** and involved the development of high-yielding varieties of cereals **[1]** (rice and wheat) and new farming techniques **[1]** including the use of irrigation **[1]**, synthetic fertilisers **[1]** and pesticides **[1]**. **[Up to 4 marks]**
4. **Any suitable answer, e.g.** The Organopónicos in Havana, Cuba, is a good example of a local scheme to increase food security among the urban poor **[1]**. It was introduced after the collapse of the Soviet Union, when Cuba lost its ability to produce enough food for its population **[1]**. It is an example of urban farming that is a sustainable practice, allowing urban dwellers to produce enough food for their families **[1]**, and can include a variety of activities such as vegetable and fruit growing **[1]**. Organopónicos take up 3.4% of urban land country-wide, and 8% of land in Havana **[1]**. They produce over 3 million tonnes of organic food **[1]** and calorie food intake is back at 2600 calories a day after the threat of hunger was a possibility **[1]**. **[Up to 5 marks]**

Page 118: Water 1

1. When there is not enough water to meet all demands as a result of physical conditions. **[1]**
2. A lack of investment in water systems or insufficient human capacity to meet the demand for water in areas where the population cannot afford to use an adequate source of water. **[2]**
3. **Any suitable answer, e.g.** LICs can improve water security by reducing wastage and evaporation **[1]**, transferring water from regions of surplus to regions of deficit **[1]** and encouraging small-scale schemes at a local level **[1]**. **[Up to 2 marks]**
4. **Any suitable answer, e.g.** Climate change is altering patterns of rainfall around the world **[1]**, causing shortages and droughts in some areas **[1]**, leading to desertification **[1]**. In the future, over half of the world's population may face water shortages **[1]**. **[Up to 3 marks]**

Page 119: Water 2

1. Water held underground in the soil or in spaces in rock. **[1]**
2. Diverting supplies from one river system to another river **[1]**; the building of dams and reservoirs increases the supply of water **[1]**; a number of countries have opened desalination plants **[1]**.
3. Groundwater is often found in porous rocks deep underground in reservoirs called aquifers. Groundwater is relied on in many developing countries, especially in Africa, because it can often be found close to villages **[1]**. Aquifers are often slow

to recharge (fill up) **[1]**, so may not always be sustainable, and groundwater supplies can become contaminated **[1]**.

4. **Any suitable answer, e.g.** The Agra clean water project, India. Agra is a city of 1.3 million people **[1]**. The city's water supply is dependent on the Yamuna River, which provides a limited supply of polluted, undrinkable water **[1]**. Those who can afford it purchase bottled water or household filters **[1]**, while people living in one of the city's 432 slums generally either tap groundwater supplies or depend on private tankers, which bring in water from outside of the city **[1]**. The project covers two areas of slums in Agra **[1]**. They are not connected to the water network and the groundwater is highly polluted by local industry **[1]**. The Agra clean water project will combine water testing and an education programme **[1]** to revive traditional knowledge and systems of rainwater harvesting and water conservation **[1]** to ensure that around 2500 people have their water supplies improved **[1]**. **[Up to 5 marks]**

Page 119: Energy 1

1. The uninterrupted availability of energy sources at an affordable price. **[1]**
2. A lack of fuel resources available **[1]** and a reliance on imported fossil fuels **[1]**.
3. **Any suitable answer, e.g.** Agriculture uses oil products to power farm machinery **[1]**, for transport of goods and livestock **[1]**, and in agricultural chemicals such as fertilisers and pesticides **[1]**. In recent years, agricultural goods like barley, maize and sugar cane have been used to make biofuels **[1]** that are used as a substitute for oil-based fuels **[1]**. Rising prices for oil mean a higher price for biofuels and agricultural chemicals **[1]**, making food more expensive **[1]**.
4. **Any suitable answer, e.g.** Demand for fossil fuels, especially oil and gas, may mean that producers have to exploit areas that are challenging **[1]** such as in deep oceans **[1]**, polar regions **[1]** and other remote, difficult and environmentally sensitive areas. The development of technology has enabled this **[1]**.
(Marks are also available for examples, such as, in the 1960s British companies searched for oil and gas in rocks underneath the North Sea **[1]**, and, in Alaska, exploitation of oil sources has taken place in environmentally sensitive and remote areas close to the Arctic Ocean **[1]**.) **[Up to 5 marks]**

Page 119: Energy 2

1. Fuels produced directly or indirectly from organic material including plant materials and animal waste **[1]**.
2. **Any suitable answer, e.g.** Countries that are LICs may not be able to afford to build them **[1]**. If they receive low amounts of rainfall, this may cause water shortages **[1]**. If there is only flat land, valleys cannot be made into dams successfully **[1]**. Dams cause environmental damage to rivers **[1]**. **[Up to 3 marks]**
3. Geothermal is a sustainable source of energy using heat from the Earth **[1]**. It is mainly exploited in areas of volcanic activity, such as Iceland **[1]**. The steam produced is used to power turbines to create electricity **[1]**.
4. **Any suitable answer with two advantages and two disadvantages, e.g.**
Advantages: wind is a renewable energy source and there are no fuel costs **[1]**; no harmful polluting gases are produced **[1]**. Disadvantages: wind farms are noisy and may spoil the view for people living near them **[1]**; the amount of electricity generated depends on the strength of the wind / if there is no wind, there is no electricity **[1]**.

Page 119: Energy 3

1. Safeguarding **[1]** and cutting down on the use **[1]** of precious supplies of energy.
2. **Any suitable answer, e.g.** Energy can be conserved in buildings with methods such as the use of insulation in loft spaces **[1]** and walls **[1]**, double-glazed windows **[1]** and larger windows in south-facing walls **[1]**. **[Up to 3 marks]**
3. **Any suitable answer, e.g.** Transport in cities can be made more sustainable by encouraging people to use bicycles rather than cars **[1]**, providing bike lanes **[1]** and offering bikes for hire **[1]**. Other examples include the congestion charge in central London **[1]**, park-and-ride schemes **[1]**, tram networks **[1]** and hybrid buses **[1]**. **[Up to 5 marks]**
4. Technology that enables coal-fired power stations to be efficient in the use of fossil fuels **[1]** and reduce the amount of greenhouse gases produced **[1]**.

Page 120: Fieldwork

For all questions, you will need to know the titles and be able to describe the location for both of your enquiries. The answers given below are suggestions for things you may consider depending on the nature of your enquiries.

1.
- For quantitative data: scatter graphs, line charts, pie charts, bar charts, histograms with equal class intervals, particular types of bar charts such as population pyramids, divided and cumulative bar and line charts, pictograms, proportional symbols, choropleth maps, isolines, heat maps, dot maps, desire lines and flow-lines. You could also include tools for statistical analysis such as lines of best fit; measures that show the strength of correlation, such as R coefficients; Spearman's Rank correlation; measures of central tendency: median, mean, mode and modal class; measures of spread and cumulative frequency: range, quartiles and inter-quartile range, dispersion graphs; percentages: percentage increase or decrease, percentiles; relationships in bivariate data: trend lines through scatter plots, lines of best fit, positive and negative correlation, strength of correlation; predictions and trends: interpolation, extrapolation; limitations and weaknesses in selective statistical presentation of data; geo-spatial (geolocated or georeferenced) data presented in a geographical information system (GIS) framework; GIS can also be used to analyse spatial data.
- For qualitative data: annotated images, diagrams or field sketches, overlays using GIS or tracing paper, tables comparing quotes or opinions.
[Up to 3 marks]
2. For the technique used, briefly explain what the presentation technique is and how it shows or enhances the information provided by the data. **[Up to 4 marks]**
3. **Structure your answer, considering the following:**
Potential risks and likelihood: slipping (high), clothing becoming wet, footwear leaking (high), hypothermia (medium – depends on weather), drowning (water not deep so low risk), allergies, traffic / wildlife (low), local conditions, e.g. stability of slopes. Who may be affected: students, teachers and other accompanying adults, members of the public. **[Up to 4 marks]**
4. **A good answer should consider the following:**
What are the potential risks? Who may be affected by the risks? What is the level or likelihood of the risk (high, medium, low)? What can be done to reduce / manage the risk? If these measures are taken, how likely is the risk to happen then? **Here is an example of a student's notes for writing a risk assessment:**
Potential risk: water in river, nearby streams, traffic near coach;
People affected by the risks: students / staff falling in the water, becoming dangerously cold, slipping or falling on rocks;
Level / likelihood: medium for most, traffic – low;
Risk reduction / management: appropriate clothing and footwear, discussion and advice about risks beforehand, reminders at key times;
Impact of measures: reduced likelihood; if something happens, everyone is prepared and knows what to do. **[Up to 4 marks]**

Pages 121–127 Mixed Questions

1. **Any suitable answer, e.g.** The USA and France are HICs **[1]** and use large amounts of water for agriculture, industry and domestic purposes **[1]**. China and India are NEEs **[1]** and their water usage is increasing due to industrialisation **[1]** and increasing domestic demand **[1]**. Mali and Egypt are both arid countries **[1]** but Mali is a LIC whereas Egypt is a NEE **[1]** and uses lots of water for irrigation purposes, having plentiful supplies (River Nile) **[1]**. **[Up to 5 marks]**
2. An extreme natural event (e.g. earthquake, volcano, tropical storm) **[1]** that causes loss of life and / or severe damage to property or severe disruption to human activities **[1]**.
3. **Any suitable impacts occurring immediately, e.g.** Falling glass; falling masonry; loss of power supplies. **[2]**
4. a) False **[1]** b) True **[1]** c) True **[1]** d) True **[1]** e) True **[1]**
5. The Saffir–Simpson scale classifies a tropical storm into one of five categories **[1]** based on its sustained wind speed **[1]**.
6. The UK's location in the mid-latitudes on the coast of north-west Europe beside the Atlantic means that there is frequent

conflict between cold and warm air masses [1] creating fronts, which produce precipitation [1].

7. hockey stick [1]

8. **Any suitable answer, e.g.** Too many beetles born in one year means more food for the birds that prey on them, so the beetle population is reduced to a sustainable level. [2]

9. **Any suitable answer which gives the idea that replenishment is needed as there are always processes at work which remove surface debris from beaches, e.g.** Beaches can be subject to longshore drift, which removes material [1] along the shore. Material is taken offshore by destructive waves [1]. Unless more material is added from weathered and eroded cliffs [1] or longshore drift [1], the beach will disappear. **[Up to 3 marks – maximum of 2 marks if only removal process(es) described]**

10. Taiga [1]

11. Salinisation [1]

12. **Any suitable answers, e.g.** (Heavily) polluted/toxic land; derelict, industrial sites [1]

13. **Any suitable answers, e.g.** Tree-labelling schemes [1]; debt reduction [1]

14. Causes: Intense low pressure in a tropical storm creates a dome of seawater, especially around the eye of the storm. As this moves across land, floods occur. [2]
Potential impacts: coastal floods; infrastructure damaged; drownings; if sewage becomes mixed with floodwater, disease can spread. [2]

15. Rural-to-urban migration of agricultural workers [1]

16. Frost-shattering [1]

17. **Any suitable answer, e.g.** Steep slopes make roads or railways difficult / impossible to build [1]. Troughs form good routeways [1] but building may need to be on the sides [1] to guard against flooding on the flat floor [1]. **[Up to 2 marks]**

18. **Any suitable impacts occurring later on, e.g.** Loss of jobs; polluted water supplies; crime (looting) [2]

19. **Any suitable answer, e.g.** Installing solar panels [1]; using energy-efficient devices like LED light bulbs [1] and plastic kettles [1]; turning off appliances when not in use (rather than using standby buttons) [1]

20. **Any suitable answer, e.g.** The use of showers and low-flush toilets [1] along with 'green' appliances that use little water [1]. Greywater harvesting [1] is a way of conserving water involving the recycling of water used in baths and showers [1] as well as from rainwater from roofs to use for flushing toilets and other non-drinking purposes [1]. **[Up to 4 marks]**

21. **Any suitable answer, e.g.** Costs involved in purchasing the technology; fuel; availability of replacement parts and maintenance [2]

22. **Any suitable answer with three points (at least one action and one result), e.g.** Particles held in the water [1] are used as tools [1] to scrape / scour the channel bed and banks [1]. The channel is cut downwards [1] and can be widened [1], especially on bends. **[Up to 3 marks]**

23. a) Notch [1]
    b) **Most likely answers are salt crystal growth or frost-shattering (or plant root and / or animal burrowing):**
    Salt crystal: salt grains from sea spray / seawater dries in cracks / joints [1] and expands when wet again to put pressure on rocks [1].
    Frost-shattering: water in cracks / joints expands on freezing [1] and puts sufficient pressure on rocks to break them [1].
    Plant roots / animal burrowing: plant roots and / or animal movement widens cracks, joints and bedding planes [1], causing fragments of rock to break away [1].
    c) Rockfall (or 'fall') [1]

24. **Any suitable answer, e.g.** Calorie intake varies between rich and poor countries. People in the USA and Europe, with over 3000 per day [1], consume the most calories per capita on average [1] and this is much higher than the recommended daily calorie intake of 2500 for men and 2000 for women [1], leading to obesity in these countries. Some African countries are experiencing below average per capita calorie intake, with averages of 2300 calories per day [1], while South Asia has an average of around 2700 [1]. **[Up to 3 marks]**

25. A greenfield site is one that has never been used for development before [1]; a brownfield site has had a previous (now redundant) industrial use [1].

26. Previous tributary river valleys have lost their confluence [1] due to erosion of the spurs into truncated spurs [1], so water tumbles over the edge.

27. **Any suitable answer – key elements are:** line of most efficient flow; erosion on outer bend(s); deposition on inner bend(s); pushing out of a bend into a definite meander.

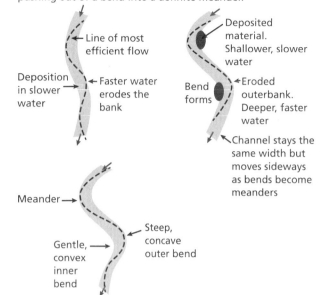

[1 for diagrams; 3 for annotations / labels]

28. 68% [1]

29. **Any suitable answers, e.g.** Regeneration seen as a 'wish list', not a reality for poorer and ethnic minority groups; new housing is expensive and attracts non-locals; fewer affordable houses provided; some local businesses forced out; jobs in retail likely to be low pay; other employment opportunities in tech and finance / international companies attract migrants from overseas to higher paid posts [3]

30. Life expectancy, literacy, and income **[2 if all correct; 1 if two correct]**

31. Greater use of nuclear [1] and renewables [1] to generate electricity.

32. **Any suitable answer, e.g.** Ethical consumerism is the purchase of sustainable goods [1], such as sources of food that are locally grown [1], and only eating fruit and vegetables that are in season [1]. **[Up to 2 marks]**

33. **Any suitable answer with four linked points (it should mention at least one specific erosion process and debris removal, and there may be reference to back-to-back caves), e.g.** Waves weaken the back of a cave using both hydraulic action [1] and abrasion [1], causing the rock to break away [1] and remove loose material [1]. At high tides, the roof of a cave will be eroded [1]. Eventually, the back of the cave breaks through to the other side of its headland [1].

34. **Any suitable answer, e.g.** Carbon footprints in HICs tend to be higher [1] due to the lifestyles people lead, as they have a greater reliance on fossil fuels [1] and they tend to use electrical devices in the home and at work [1]. People in HICs have a diet that tends to rely on food that is imported [1] or is grown using high levels of inputs derived from fossil fuels [1]. In LICs, carbon footprints are generally lower [1]. Lifestyles are more sustainable because lower amounts of fossil fuels are used [1] and food tends to be locally produced [1]. **[Up to 6 marks]**

35. **Answers will vary but should consider these points:**
Reliability of conclusions: For the topic of your enquiry, you need to make judgements about how close your conclusions are to actual changes happening.
Comments on reliability may be affected by limitations of equipment used; equipment operator errors; choice of data collected; methods of data collection; design methodology / sampling methodology, e.g. number and suitability of sample sites (spatial) and time of day or year (temporal).
Use these ideas to reach an overall judgement about the reliability of your conclusions. Also evaluate the extent to which your findings would be repeated if the investigation was undertaken at different places (spatial) or times (temporal).
[9 + 3 SPaG]

Glossary and Index

3 Ps: Prediction; Protection; Planning measures a community or country can take to improve their resilience to a hazard such as a tropical storm **9, 17**

abrasion erosion process; use of material carried in water (sea, river) to scour and scrape **45, 51, 56–57**

active volcano a volcano with at least one eruption in the last 10 000 years **12**

adaptation measures taken to reduce the negative impacts, or take advantage of the positive impacts of, climate change **21, 30–31, 36–37, 107**

aesthetic a sense of beauty or attractiveness **60**

air mass a large body of air with the same temperature throughout **18–19, 28**

alluvium material deposited by rivers **53**

altitude height above sea level **18, 60**

angular sharp-edged shape of rocks, not rounded at all **56**

anthropogenic caused or produced by humans **21**

appropriate technology science or technology suitable for the area **35, 79, 96**

aquifer an underground layer of rock that bears water **35, 100–101**

arable the farming of crops **60–61**

arch erosion feature; headland that has been cut through away from the seaward end to form a bridge shape **46**

arête steep-sided, narrow, rocky ridge formed between corries **58–59**

arid region an area with a severe lack of available water **98–99**

asymmetrical not identical either side of the middle; in rivers, a channel almost always has one bank steeper than the other **6, 59**

atmospheric circulation the movement of air around the Earth in cells, transferring and redistributing energy **14, 28**

atmospheric pressure the pressure exerted by the weight of the atmosphere **28**

attrition process in which particles get smaller and rounder as they crash into each other in water (sea, river) **45, 51**

backwash return wave motion to the sea **44–45**

bar strip of deposited sediment parallel to the coast **47**

beach deposition feature; loose material, usually sand or shingle **47**

bedload material on the river channel floor; generally the biggest particles and most difficult to move **51**

biofuel fuel produced from renewable resources such as vegetable oils **103, 104**

biomass renewable material from living things **104**

blocking high slow-moving anticyclone (high-pressure system) **18**

brownfield site land previously used for industrial purposes **73, 85**

buttress roots a shallow root system ensuring the quick take-up of nutrients **30**

carbon capture (carbon sequestration) the removal of carbon at emission source and storage of it deep underground **21**

carbon footprint the amount of carbon dioxide released as a result of an organisation's/individual's activity **92, 106**

carbon neutral taking action to remove as much carbon dioxide from the atmosphere as is put into it **74**

carnivore an organism, such as an owl, that eats other animals **26–27**

channel the groove in which water flows **50–51, 52–53, 54–55**

channel catch rainwater falling directly into the river **54–55**

choropleth map uses shading to indicate average values **7**

clean coal technology seeking to reduce harmful coal-based emissions by using technology **104, 107**

climate average weather pattern over time; expected conditions **4, 18, 29, 30–31, 60**

collision zone a plate margin where two tectonic plates crash into each other **9**

colonialism where one country goes into another country and claims they are in power, often exploiting people and resources **78**

commercial farming crops are produced on a large-scale for profit **34**

composite volcano cone-shaped volcano made up of many layers **12**

confluence where any two rivers meet **46, 56–57**

conservative margin formed when plates slide past each other **8**

constructive margin formed when plates move apart **8**

constructive wave low energy waves which add material to a beach; more effective swash than backwash **50**

consumer a species that eats another species **26–27**

continental coming from a nearby continent **18**

contour a line on a map along which the height is the same **5, 7**

core the central region of the Earth **8–9**

Coriolis force (Coriolis effect) the result of Earth's rotation on weather patterns **15**

corrie curved rock hollow in a mountainside with a steep back and ice-carved base; may contain a lake **58–59, 60**

counterurbanisation where people move from urban areas to rural areas **83**

critical thinking the ability to interpret, analyse and evaluate ideas and arguments **110–111**

crust the topmost layer of the Earth, made up of tectonic plates **8**

cumec cubic metres per second; the measurement of discharge in a river **54**

cyclone a tropical storm in the Indian Ocean and South Pacific, e.g. India **14**

dairy the production of milk **60**

debris loose material that has been weathered or worn away from a landscape and can be carried away or deposited **46, 56–57, 58–59**

decompose to break down **44**

decomposer an organism, such as fungus, that breaks down the dead remains of another organism **27**

de-industrialisation a decline in the number of workers or the output in manufacturing industry **82**

Demographic Transition Model shows birth rate, death rate and population change over time **77**

dense pattern or distribution is dense if it shows a high concentration in one area, e.g. dense drainage pattern; dense population **4**

deposition material being left behind in response to a loss of energy **44, 45, 51, 52–53, 57, 59**

desalination removing minerals from saltwater to render the water suitable for human use **100**

desertification the process by which previously fertile land becomes desert **34–35**

destructive margin a plate boundary that occurs when plates move together **8**

destructive wave high energy waves which remove material from a beach; backwash is more effective than swash **44**

dispersed a type of settlement pattern: separate and scattered **4**

diversification use of farm land or buildings for non-agricultural activities **61**

dormant volcano an active volcano that is not currently erupting **12**

downcutting vertical erosion of a channel, making it deeper **51**

drainage basin water from an area that feeds into a river; also known as a river basin or a catchment area **50**

dredging scraping material out of a channel to make it bigger **55**

drought a period of below average precipitation **18–19, 95**

drumlin elongated glacial deposit with a blunt upstream side (stoss end) and tapered downstream side **59**

easting line one of the vertical lines crossing an OS map from top to bottom; they are called eastings as the numbers increase in an easterly direction **4**

economic development improvement in living standards by creation of jobs **32, 98, 103**

economic leakage when money that is generated in a country leaves a country; this happens when a transnational corporation sends money back to its home country **80**

economic water scarcity the population does not have the monetary means to access available water **98**

ecosystem a community of interdependent living and non-living elements that create a particular environment **26–27, 28–29, 30–31**

ecotourism responsible travel to areas that sustains the well-being of the local population and environment **33**

elevation profile shows how far above sea level certain features are **5**

embankment artificial bank built alongside a channel; can be of natural materials or concrete **55**

energy consumption amount of energy used by individuals or groups **102–103**

energy insecurity without access to a secure and affordable energy supply **102–103**

energy poverty unable to heat or provide other energy services to homes **102–103**

energy security uninterrupted availability of energy sources at an affordable price **93, 102**

en-glacial (moraine) debris inside the ice **57**

enterprise zones areas which provide incentives to attract business investment **83**

entrained material being taken into an agent of transport and erosion, such as a glacier or river **50**

environmental relating to the natural world and the impact of human activity on its condition **60, 72, 73**

epicentre the point on the Earth directly above the focus of an earthquake **10–11**

erosion wearing away of the landscape due to the movement of ice, water or wind **45, 46, 50–51, 56–57**

erratic a rock or boulder that differs from the surrounding rock **59**

estuary tidal river mouth where saltwater from the sea meets freshwater **47, 53**

ethical consumerism buying only products or services that are produced in a way that does not harm the environment **97**

extinct volcano a volcano that has not shown an eruption for at least 10 000 years, and is not expected to erupt again in the future **12**

extrapolation predicting future outcomes based on known facts **7**

extreme temperatures excessive heat or cold **31, 34**

extreme tourism travel to remote or unsettled areas which can be dangerous **37**

eye of the storm if a tropical storm becomes 'cyclonic', it spins so fast that the air around the centre forms a vortex which has an eye 20–40 miles (30–65 km) wide **15**

fair trade farmers paid a fair price for their goods **79**

famine extreme scarcity of food **93**

faulting breaks in rocks; can be as small as a hairline or huge **52**

favela an area of makeshift housing in or near a city in Brazil **70–71**

fetch distance over which a wave has travelled **44**

flashy a river that responds quickly to rainstorms **54**

flocculate to cause material to join together, gain weight and sink to the bottom of a river/sea **53**

flood a river going over its bank **51, 53, 54–55, 60**

floodplain low-lying area made of deposited material to sides of a river channel; flood waters contribute to the formation **51, 53**

fluvial the processes associated with rivers and streams **50**

flood management measures which are taken to alleviate the negative impacts of flooding on people and their property **6, 54–55**

fodder coarse food for livestock, e.g. hay, straw **60**

food chain simple set of connections showing how a small group of organisms are linked; who eats what **26**

food insecurity being without access to enough food **94–95, 96–97**

food miles the journey and fuel used from food producer to consumer **75, 92**

food security having reliable access to enough food **94–95, 96–97**

food web diagram showing how larger groups of organisms are interlinked **26–27**

fossil fuel sources of energy formed from the remains of organisms buried millions of years ago; coal, oil and gas **102–103, 104–105, 106–107**

fracking drilling the earth and injecting high pressure liquid to extract gas **93**

freeze-thaw   breaking down rocks by frost shattering, that is water entering cracks, expanding when frozen and so putting pressure on the rock and weakening it   56, 58

frequency   regularity of an event such as a tropical storm   7, 15

friction   force between surfaces trying to pass each other; in a river, there is friction between the water and channel   8, 52

front   a boundary separating two masses of air of different densities   18

gabion   a bundle of rock in a wire mesh cage; used to protect slopes, especially at the coast, on river banks, and roads   48

geographical enquiry   investigation linking your study in school/college with fieldwork carried out away from school/college   108

geology   the science of Earth's origins, including rocks and minerals   44, 54

geothermal energy   using heat from deep within the ground (such as near volcanoes) to obtain energy   9, 105

GIS   systems designed to capture, store, manipulate, analyse, manage and present all types of spatial or geographical data; GIS data is geolocated or georeferenced using widely recognised locational methods such as latitude and longitude   5, 7, 109, 110

glacial   a cold period in ice ages when glaciers and ice sheets grow   20

glacial trough   deep, steep-sided valley formed by glaciers   58–59

glacier   slow moving stream of ice formed from accumulation of snow   56–57, 58

globalisation   process whereby the world is becoming increasingly interconnected thanks to mass communications   82

gorge   steep-sided section of a valley; usually rocky   52

gradient   steepness of a slope   47

greenfield site   a plot of land that has not been used before for building development   85

greenhouse effect   greenhouse gases stop heat escaping from the Earth into space; an enhanced greenhouse effect can lead to climate change and global warming   20–21, 104

greenhouse gas   gas in the atmosphere which absorbs and emits heat (infrared) radiation, such as carbon dioxide, methane and water vapour   20–21

greywater   waste water from the household that can be recycled   98, 101

ground moraine   thick layer of clay containing rock fragments from under a glacier; also known as till   59

groundwater   water held underground in the soil or rock   98–99, 100–101

groundwater flow   slow, underground, horizontal movement of water towards the channel   54

growing season   a period of the year when crops grow successfully   60

groyne   man-made structure at right angles to the coast to stop material travelling along the beach by the action of longshore drift   48–49

Gulf Stream   a powerful warm ocean current   18

hanging valley   tributary valley along the side of a glaciated valley   58

hard river engineering   range of measures used to manage river flow and floods which use artificial structures made from materials such as concrete, stone or steel, e.g. wing-dykes; dams; weirs; often more expensive than soft engineering and considered to be less economically, socially and environmentally sustainable   6, 55

headward   towards the source of a river   50

helicoidal   like a flattened curve; water takes this motion at meander outer bends, hitting the bank high then moving down and outwards towards the next inner bend   52–53

herbivore   an organism such as a sheep that eats only plants   26–27

HIC   higher income country   10–11, 68, 76–77

histogram   data grouped into ranges and plotted as bars   7

hot spot   where strong upward currents in the magma reach close to the surface at points on the crust away from plate margins; volcanic activity is common here   9

Human Development Index (HDI)   a measure of development taking account of several development indicators   76

hunter-gatherer   a tribesperson who lives by hunting and harvesting wild food   32

hurricane   a tropical storm with heavy rain and strong winds   14–15

hybrid vehicle   one which uses two or more types of power, such as internal combustion engine plus electric motor   107

hydraulic action   erosion process; action of moving water in rivers; air trapped in cracks etc. by waves exerts great pressure   45, 50–51, 52

hydroelectric power (HEP)   producing electricity using water power   32, 34, 105

hydrograph   diagram showing the amount of water in a particular river over time; can be produced for minutes, hours, days, months, years   54

hydro-meteorological hazard   a weather hazard linked to the water cycle   18

hypothesis   a statement or theory that can be tested   108–109

impact   the economic, social and environmental effect of an action or decision   19, 21, 111

impermeable   surface that does not allow water to pass through   6, 54–55, 60–61

industry   economic activity not usually associated with agriculture   60, 79, 84

inertia   when people and/or organisations do not take action when faced with the risk of a hazard   17

infrastructure   facilities and supply lines which make modern life possible, such as roads, railways, air and sea ports, water pipes, sewage pipes, electricity cables, telecommunications links, computer networks and Internet access   17, 34–35, 84

intensity   severity or strength of an event such as a tropical storm   15

intercept   get in the way of; anything that stops rainwater reaching the ground, permanently or temporarily   55

interest group   organisation of people with a common cause about which they try to influence policies or decisions, without seeking political control   110–111

interglacial   a period of milder climate between two glacial periods   20

interlocking spurs   fingers of land around which rivers flow; they obstruct the view up or down the valley   52

intermediate technology   technology appropriate for the developing needs of the community   79

interpolation   predicting values from a limited number of sample data points   7

interrelationship   two-way links between different elements in an ecosystem   26

IPCC   the Intergovernmental Panel on Climate Change   20–21

irrigation   artificial addition of water to land to help crops grow   30–31, 34–35, 96–97

isobar   a line on a weather map joining areas of the same air pressure   6

isohyet   a line on a weather map joining areas of the same rainfall   6

isoline   a line drawn to link different places that share a common value   5, 6–7

key player   a person or group with an interest and a significant degree of control about an issue or in the development of a decision or policy   110–111

lag time   difference in time between peak rainfall and peak discharge   54

lahar   a mud flow consisting of volcanic ash and water that runs down the slopes of a volcano, sometimes burying settlements in its path   12–13

landlocked   surrounded by land and having no navigable route to the sea   78

landward   facing away from the sea   49

lateral (moraine)   material carried at the side of a glacier and deposited along the side of a valley   57, 59

latitude   latitude lines provide a measure of how far places are north or south of the Equator; latitude lines are all parallel to the Equator   4

latosol   soil found under tropical rainforests   30

leaf litter   dead plant material on the ground surface   30–31

levée   ridge of deposited material alongside a channel; results from floods; artificial levées can be made as flood defences   53

LIC   lower income country   11, 68, 76–77

limestone   sedimentary carbonate rock made from the bodies of sea creatures   51

limitation   disadvantage or cause of potential error or the collection of unreliable or unrepresentative data   109

linear   pattern or distribution is linear if there is a concentration of something in a line or ribbon, e.g. a linear settlement pattern   4

load   material carried by an agent of erosion and transport, e.g. river, glacier   51

longitude   longitude lines provide a measure of how far places are east or west from the Prime Meridian (sometimes called the Greenwich Meridian); longitude lines run between the north and south poles at right angles to the latitude lines   4

longshore drift   movement of material along the shore due to wave action   45, 47, 48–49

long-term aid   providing help for education and skills for development   79

lyme grass   salt-loving plant; one of the first to grow on sand dunes   49

management   the forecasting, planning, organising and decision-making used by people to improve people's lives   111

mantle   the region within the Earth between the core and the crust   8–9

maritime   related to the sea   18

marram grass   tough, long rooted, wide bladed grass that lives in sand and holds the grains together   49

mass movement   downslope movement of surface debris under gravity   44–45

mass tourism   where tens of thousands of people visit the same place at the same time of year   35

mean   the sum of the quantities divided by the number of quantities   7

meander   winding curve or bend in a river   52–53

meander belt   area of a drainage basin where meanders are formed   50–51

medial (moraine)   merging of two lateral moraines   57

median   the point in a series that is exactly in the middle of the values   7

megacity   a city with over 10 million inhabitants   68

meltwater   water resulting from melting snow and ice   57

mid-latitudes   the temperate zones between the tropics and polar regions   18

millionaire city   a city with over a million inhabitants   68

mitigation   measures to reduce the causes of climate change, e.g. cutting greenhouse gas emissions; eliminating long-term risk to human life and property; international targets to reduce greenhouse gas emissions; carbon capture   21

modal class   the class with the highest frequency   7

mode   the most frequently occurring data value in a series   7

moraine   material carried and deposited by ice   57

mouth   where a river ends; usually meeting the sea   50, 53

mudflats   low-lying area of silt, clay and sand particles found in low energy environments, exposed at low tide   53

NEE   newly emerging economy   68, 70, 76, 80

negative multiplier effect   where a negative event triggers other negative events   83

North Atlantic Drift   a powerful warm ocean current   18

northing line   one of the horizontal lines crossing an OS map from one side to the other; they are called northings as the numbers increase to the north   4

notch   small indentation in a cliff face at high tide mark; results from wave action   46

nucleated   a pattern or distribution is nucleated if there is a concentration of something in a tightly-confined cluster, e.g. a nucleated settlement pattern   4

nutrient cycle   how the minerals that provide energy and sustenance to living organisms circulate around an ecosystem; minerals are moved between the atmosphere, biomass, litter layer and soil   27

offshore   away from the land   47, 48–49

organic   produced without the use of artificial chemicals   92, 97

orientation   direction, using compass points or bearings   5

out-migration   moving away from a local community   33

overland flow   water moving over the surface towards the channel   54–55

ox-bow lake   formed when a river meander is cut off   53

pastoral farming   farming aimed at producing livestock rather than crops   35, 60

percentile   each of the 100 equal groups into which a population can be divided according to the distribution of values of a particular variable   7

peat   partly decomposed vegetation and organic matter; holds a lot of water   55

peripheral   describes places which are distant from the centre or core of activity   4

permaculture   the development of agricultural ecosystems intended to be sustainable and self-sufficient   97

permafrost   land that has been frozen for at least two consecutive years   36

permeable   a rock or surface that allows water to pass through   6

physical water scarcity   a situation where the water is not abundant enough to meet all demands for the population   98

plate margin   the boundary of the plates that form the upper layer of the Earth   8

plateau   flat, elevated land rising sharply above the area surrounding it on at least one side   5, 52

plucking   erosion process; weakened rock removed from bedrock by a glacier   **57, 58**

plunging   destructive waves that bring material from a beach towards the sea   **44**

political map   shows boundaries linked to areas of governance such as countries, states, and counties, and the location of major cities   **4**

pollution   the introduction by humans of harmful material into the environment; always specify what type of pollution: air pollution; water pollution, etc.   **61**

prevailing wind   a wind from the direction that is most usual at a particular place or season   **18**

primary data   data that you and your fellow students have collected   **109**

primary effects   the effects that occur immediately after the natural disaster happens, e.g. the ground shakes   **10–11, 12–13**

problem-solving   thinking about the possible solutions to a problem and deciding which course of action to take   **110**

producer   an organism, e.g. a tree, that makes its own food via photosynthesis   **26–27**

proxy measure   an indirect way of measuring variables, such as tree ring growth as an indicator of temperature in past years   **20**

pull factors   positive factors attracting people to live in a place   **69**

push factors   negative factors making people want to leave a place   **69**

pyramidal peak   sharp-pointed, frost-shattered mountain top with corries on at least three sides   **58**

pyroclastic flow   a high temperature avalanche of gas, ash, cinder and rock that rushes down the slopes of a volcano at speeds of up to 450 mph   **12–13**

qualitative data   non-numerical data which might involve subjective judgements, e.g. field sketches; photographs; video; quotes or opinion   **109**

quality of life   the wide range of human needs that should be met alongside income growth such as access to housing, education and health, nutrition, security, happiness, etc.   **76**

quantitative data   numerical data or statistics, e.g. width and depth of a river channel; an environmental quality survey   **109**

quartile   one of four equal groups into which a population can be divided according to the distribution of values of a particular variable   **7**

Quaternary period   the last 2.6 million years in which there have been several glacials   **20**

radial   pattern spread out in a linear way from a central point, like the spokes on a bicycle wheel or rays of the sun, e.g. rivers or glaciers flowing in all directions from high ground; transport links from a hub   **4, 6**

rain shadow   a dry area on the leeward side of a hilly/mountainous area   **18**

regeneration   reviving old run-down urban areas by either improving what is there or clearing it and rebuilding   **73**

relief   the difference in elevation in an area; shape of the land   **4, 6, 60**

relief map   shows the height and shape of the land (topography) using symbols such as contours or layer shading   **4**

relief rainfall   occurs when moist air rises over a physical barrier, e.g. mountains   **18**

renewables   a natural resource and source of energy that is not depleted by use, such as water, wind or solar power   **93**

reprofiling   change the face/front of a slope; could be to make more or less steep, higher or lower, than originally   **49**

resilience   capacity to cope with a hazard such as a tropical storm, often dependent on how effectively the 3 Ps are implemented   **17**

resistant   tough; able to withstand weathering and/or erosion   **44**

ribbon lake   long narrow lake formed by a glacier   **58–59, 60–61**

Richter scale   a measure of the energy released by an earthquake   **10–11**

ridge   mountains forming a continuous elevated crest   **5**

river bank   the land alongside a river, above the channel   **51**

salinisation   the increase in the salt content of water   **31, 35**

saltation   bouncing or jumping movement of particles along a surface   **45**

saltmarsh   area of deposition of fine material that is tidal; adapted plants grow and stabilise the area   **49**

sampling techniques   data is collected from a small part of the population or a small number of sites and used to inform what the whole picture is like; there are three main types of sampling: random; systematic; stratified   **109**

sand dune   accumulation of sand in hill form at the back of beaches along low-lying coastlines   **47**

saturated   no more water can be held; if a permeable material becomes saturated, it becomes impermeable   **54**

scale   ratio which compares a measurement on a map between places to the actual distance between these places in reality; for example, the most commonly used OS maps use a scale of 1:50000 and 1:25000; scale is also used when referring to places or areas of different sizes, e.g. local, regional, national, international, global scale   **5**

sea couch   salt-loving plant; one of the first to grow on sand dunes   **49**

secondary data   data collected by another person, group or organisation which you have accessed via articles, websites, books, etc.   **109**

secondary effects   the effects that occur subsequent to the primary effects, such as gas mains rupturing   **10–11, 12**

settlement   a place where people establish a community   **4, 17, 32, 34, 37, 70–71, 72, 75**

shield volcano   large sized, low-lying volcano formed from fluid magma flows   **12**

shifting cultivation   ground is cultivated and then abandoned to allow its fertility to be restored   **32**

short-term aid   immediate relief after an emergency   **79**

silage   crops set aside for farm animals in winter   **60**

sinuous   curving movements in a river's flow   **53**

slip-off slope   area of deposition at the inside bend of a meander   **53**

soft river engineering   range of measures used to manage river flow and floods which do not use materials such as concrete, stone or steel, e.g. river restoration, floodplain zoning; they allow natural processes to occur and don't alter the natural environment; often less expensive than hard engineering and considered to be more economically, socially and environmentally sustainable   **6, 55**

solution   dissolving action in water   **44, 51**

source   beginning of a river   **50**

sparse   a pattern or distribution is sparse if it shows a low density and/or is very dispersed in an area, e.g. sparse drainage pattern; sparse population   **4**

spilling   wave which pushes material onshore   **44**

spine   the main channel in a river system; also known as the trunk   **50**

spit   long protrusion of deposited material extending from the coast into the sea   **47**

spot height   a point of known height shown using a dot with a number beside it, usually in metres above sea level; appears on a map, but not on the landscape   **5**

spur   a piece of land jutting into a river or stream   **5**

stakeholder   a person or group with a real interest in an issue or in the development of a decision or policy   **110–111**

steep   having a noticeable slope of over 30°   **54**

storm surge   a rising of the sea as a result of wind and atmospheric pressure changes associated with a storm   **15, 17**

subdued   description of a river which responds very slowly to rainstorms   **54**

sub-glacial (moraine)   debris under the ice   **57**

subsistence farming   crops produced solely for the farmer and family   **32**

supra-glacial (moraine)   debris on top of the ice   **57**

surging   wave which pushes material up a beach, making it steeper   **44**

suspended load   particles carried along within the body of water (river or sea)   **51**

sustainable   meeting the needs of the present population without compromising the needs of future generations   **33, 74, 85, 97, 100, 106**

swash   forward wave motion   **44**

symmetrical   similar gradient on both sides   **6**

synoptic thinking   the application of knowledge, understanding and skills from all of your geographical learning to a new situation   **110**

terminal (moraine)   crescent-shaped deposit at the furthest reach of a glacier   **57, 59**

terraced field   used for farming in hilly regions   **35**

terracing   making a slope into shallow steps; this makes the overall slope less steep and makes flat surfaces to hold water   **55**

terrain   landscape pattern   **60**

thalweg   the fastest part of a river, where there is least friction   **52**

throughflow   lateral movement of water below the surface (through the soil towards the river)   **54**

tidal   influenced by the sea; seawater advances and retreats with the tide   **53**

till   clay-containing rock fragments deposited from under a glacier   **57, 59**

tourism   activities of visitors to an area and those who cater for them   **34–35, 61, 80–81**

transnational corporation (TNC)   enterprise operating in at least two countries   **80, 102**

traction   rolling and sliding of larger, heavier particles along a river bed or in the sea   **45, 51**

transpiration   water evaporates from plants into the atmosphere   **31, 54**

transportation   material being carried away; in rivers, by solution, suspension, traction and saltation   **51, 57**

triangulation pillar   a point of very precise known height shown using a dot with a number beside it, usually in metres above sea level; appears on a map as a triangular symbol, and is marked on the landscape with a concrete pillar   **5**

tributary   any river flowing into another, bigger river   **50, 58**

tropical storm   low pressure systems with distinct structure and features formed over warm ocean waters in low latitudes   **14–15, 16–17**

trough end   upstream end of a glaciated valley with a sudden closure   **58**

truncated spurs   former interlocking spurs from a river valley which have been cut away by glacial erosion   **58, 61**

trunk   the main channel in a river system; also known as the spine   **50**

tsunami   a large wave caused by an earthquake, volcanic eruption or coastal landslide   **10–11**

tundra   vast treeless zone in the arctic regions of the world   **29, 36–37**

typhoon   tropical storms in the western North Pacific, e.g. Philippines, Japan   **14–15, 16–17**

undermine   erode from below; cause weakness   **51**

urbanisation   an increase in the number of people living in towns and cities compared with rurally   **68–69**

U-shaped valley   deep, steep-sided valley formed by glaciers; also known as a trough   **58**

valley   between areas of higher land, often with a river running through   **5, 6, 8, 58–59**

waterborne disease   caused by coming into contact with an infected water source   **99, 101**

water conservation   beneficial reduction of water usage   **100**

water deficit   not enough rainfall, leaving a water shortage   **19**

waterfall   where river water flows over a sudden vertical break   **52, 58**

waterlogged   saturation of the soil by an excess of water   **55, 60**

watershed   the dividing line between two drainage basins   **50**

water security   the reliable availability of water to sustain the population   **98**

water stress   lacking water   **19**

water surplus   having more water than needed   **19**

water table   level below ground at which rocks are saturated   **47, 54**

water transfer scheme   moving water from an abundant river basin to one where water is less available   **100**

wave cut platform   gently shelving area of solid rock stretching out to sea from the cliff front   **46**

weathering   break-down of rock as a result of exposure to the atmosphere   **44, 46, 56–57, 58–59**

wetted perimeter   length of channel bed in contact with water   **50, 54**

wilderness   areas untouched by human influence   **37**

wind patterns   the semi-predictable movement of masses of air around the world, e.g. the trade winds   **28**

yield   the measure of crops produced, often in kilograms per hectare   **95, 96**

Collins

AQA GCSE 9-1
Geography

Workbook

Janet Hutson, Dan Major, Paul Berry,
Brendan Conway, Tony Grundy, Robert
Morris and Iain Palôt

Preparing for the GCSE Exam

Revision That Really Works

Experts have found that there are two techniques that help you to retain and recall information and consistently produce better results in exams compared to other revision techniques.

It really isn't rocket science either – you simply need to:

- **test yourself** on each topic as many times as possible
- **leave a gap** between the test sessions.

Three Essential Revision Tips

1. Use Your Time Wisely

- Allow yourself plenty of time.
- Try to start revising at least six months before your exams – it's more effective and less stressful.
- Don't waste time re-reading the same information over and over again – it's time-consuming and not effective!

2. Make a Plan

- Identify all the topics you need to revise (this Complete Revision & Practice book will help you).
- Plan at least five sessions for each topic.
- One hour should be ample time to test yourself on the key ideas for a topic.
- Spread out the practice sessions for each topic – the optimum time to leave between each session is about one month but, if this isn't possible, just make the gaps as big as realistically possible.

3. Test Yourself

- Methods for testing yourself include: quizzes, practice questions, flashcards, past papers, explaining a topic to someone else, etc.
- This Complete Revision & Practice book provides seven practice opportunities per topic.
- Don't worry if you get an answer wrong – provided you check what the correct answer is, you are more likely to get the same or similar questions right in future!

Visit **collins.co.uk/collinsGCSErevision** to download your free flashcards, for more information about the benefits of these techniques, and for further guidance on how to plan ahead and make them work for you.

Command Words used in Exam Questions

This table defines some of the most commonly used command words in GCSE exam questions.

Command word	Meaning
State	Write clearly and plainly
Define	Give the meaning of a term
Calculate	Work out the value of
Outline	Give a brief account or summary
Compare	Identify similarities and differences
Describe	Write what something is or appears to be
Suggest	Propose an idea or solution
Explain	Say why or how
Assess	Make an informed judgement
To what extent	Weigh up the importance or success of
Evaluate	Using evidence, weigh up both sides of an argument
Discuss	Offer key points about the different sides, or strengths and weaknesses, of an issue
Justify	Give detailed reasons for an idea

Contents

TOPIC-BASED QUESTIONS

The Challenge of Natural Hazards   148

The Living World   152

Physical Landscapes in the UK   155

Urban Issues and Challenges   159

The Changing Economic World   161

The Challenge of Resource Management   164

Geographical Applications   168

PRACTICE EXAM PAPERS

Paper 1 – Living with the physical environment   169

Paper 2 – Challenges in the human environment   191

Paper 3 – Geographical applications   213

Resources for Paper 3 – Geographical applications   225

ANSWERS   231

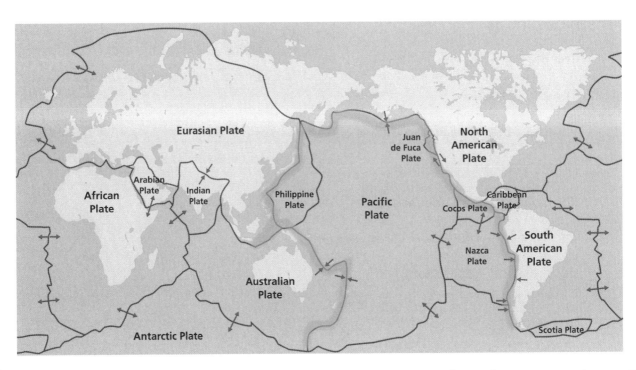

1 **a)** What name is given to the highly active area of tectonic activity shown in orange on the map? [1]

....

b) Use the map to help you name two plates that form a **destructive** margin. [1]

....

2 Explain what happens in a **collision zone**. [3]

....

....

....

....

3 What **three** items would you include in an emergency kit for people living in an earthquake zone? Explain how they might help. [6]

....

....

....

....

....

....

4 Why are **tsunamis** so dangerous? [4]

5 Describe some of the **longer term problems** which people have to deal with after an earthquake. [8]

6 Describe some of the ways that **volcanic eruptions** can be predicted. [6]

7 **a)** This map shows the tracks of **tropical storms** over the last 70 years.

Describe the pattern of tropical storm tracks in the map. [4]

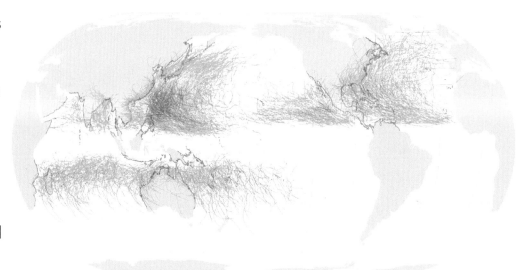

....

....

....

....

b) This map shows **average sea surface temperatures**.

Suggest reasons for the links between the distribution of tropical storms and sea surface temperatures. [4]

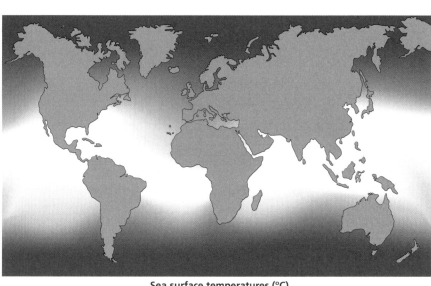

Sea surface temperatures (°C)

0   5   10   15   20   25   30

....

....

....

8 The UK has a **maritime** climate. What does this mean? [3]

9 What causes the **variations in rainfall** between the west of the UK and those regions in the east? [4]

10 How are **ice cores** used as evidence of long-term climate change? [3]

11 The **impacts of climate change** can be managed through **adaptation and mitigation**. Explain, using examples, how these two approaches might work. [8]

Total Marks _____ / 55

The Living World

1 What is **ecotourism**?   [2]

2 Large-scale ecosystems are also called '**biomes**'. Name a biome.   [1]

3 Describe **one** way in which humans can disrupt the **balance of an ecosystem**.   [2]

4 Explain how the **nutrient cycle** works.   [5]

5 Name the **two** regions where there is **low pressure in January**.   [2]

6 What measures allow tropical rainforests to be **managed sustainably**?   [6]

The Living World

7 Why do some trees in the tropical rainforest have **buttress roots**? [2]

8 What sort of trees grow in the **boreal forests**? [1]

9 How do plants that grow in desert regions **adapt** in **order to survive**? [4]

10 Define the term '**desertification**'. [1]

11 a) For **an area you have studied**, explain what has caused the problem of desertification. [4]

b) Describe what **strategies** have been adopted to deal with desertification in this area. [4]

12 **a)** As the polar regions get warmer, suggest what **resources might be exploited**. [2]

b) Suggest why this exploitation might be a **problem**. [4]

13 What is the **Madrid Protocol** and what does it prohibit? [3]

14 What is **permafrost**? [2]

15 What is **extreme tourism** and what problems might this cause? [4]

Total Marks _____ / 49

Physical Landscapes in the UK

1 What conditions have to exist for **freeze-thaw weathering** to take place? [2]

2 Explain how **abrasion** takes place under a moving glacier. [2]

3 Describe the **shape** and **formation** of **drumlins**. [4]

4 Explain why **glaciated uplands** are unsuitable for farming. [5]

5 Explain what makes glaciated uplands suitable for the **generation of hydroelectric power**. [4]

Physical Landscapes in the UK

6 Look at the photograph below showing part of a coast that is being managed.

a) Are the coastal management measures shown, **hard** or **soft** engineering?   [1]

b) State the names of the **coastal management structures** shown.   [2]

7 For a stretch of coastline you have studied, explain why **coastal protection** was necessary and what measures were put in place.   [6]

Physical Landscapes in the UK

8 Describe the conditions necessary for **spit** formation. [3]

9 Using an example you have studied, describe the advantages of '**managed retreat**'. [4]

10 Describe **four** ways by which a river can **erode** its channel. [4]

11 Explain how **meanders** are formed. [4]

12 On the diagram of a **storm hydrograph**, label the following: [7]

| Base flow | Rising limb | Lag time | Peak discharge | Falling limb | Storm flow | Peak rainfall |

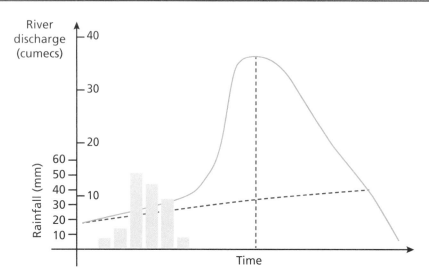

13 For a UK river that you have studied, choose **one** feature caused by **deposition** and describe its location and key features. [4]

....

....

....

....

....

14 What are the main causes of **river flooding**? [3]

....

....

....

15 Give **two** advantages and **two** disadvantages of **river hard engineering**. [2]

....

....

....

....

Total Marks .... / 57

Urban Issues and Challenges

1 Define the term 'urban sprawl'. [2]

_____

_____

2 Suggest why cities in lower income countries (LICs) are growing so rapidly. [3]

_____

_____

_____

3 Suggest **two** reasons why birth rates are higher in LICs. [2]

_____

_____

4 Give **two** push factors and **two** pull factors which result in rural-to-urban migration. [4]

_____

_____

_____

_____

5 a) For a city in a **newly emerging economy** that you have studied, describe **two challenges** it faces. [4]

_____

_____

_____

_____

b) Using the same challenges, what **two** solutions has the city implemented? [4]

_____

_____

_____

_____

6 Describe the **opportunities** created by **urban change** in a **major UK city** that you have studied. [4]

7 What **challenges** have arisen from **urban change** in a **major UK city** that you have studied? [4]

8 List some of the features of **sustainable living**. [3]

9 Describe, giving examples, **two** strategies that might help **reduce urban traffic congestion**. [4]

10 How does **urban agriculture** help to reduce food miles? [2]

11 Give **two** advantages of **urban greening**. [2]

Total Marks _____ / 38

1

Human Development Index

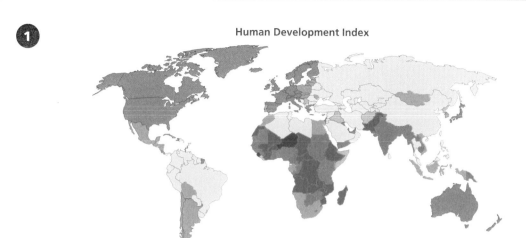

	High human development	0.9–1.00
		0.8–0.89
	Medium human development	0.7–0.79
		0.6–0.69
		0.5–0.59
	Low human development	0.4–0.49
		0.3–0.39
		0.2–0.29
	Not applicable	

Give **two** reasons why a map like this could give a wrong impression.   [2]

2 State **three** measures used to show differing rates of development.   [3]

3 What does the **Demographic Transition Model** help to show?   [2]

4 Give **two** physical causes that can lead to **global inequality**.   [2]

5 What might be **two** of the consequences of **uneven development**?   [2]

6 What is meant by the term 'intermediate technology'? [2]

7 Describe a **fair trade** that might help close the **development gap**. [4]

8 Describe how **tourism** has reduced the **development gap** in a LIC or NEE that you have studied. [4]

9 **TNCs** can play a role in **rapid economic development**. Using an example you have studied, give **two** advantages and **two** disadvantages of rapid economic development. [4]

10 The provision of aid to a LIC can have advantages and disadvantages. For a country you have studied, state a **type of aid** that has been received and describe the **benefits** to the country. [4]

11 What are the main features of the **north-south divide** in the UK?   [4]

12 Define the term '**deindustrialisation**'.   [2]

13 Describe measures being taken to make modern industry more **environmentally sustainable**.   [4]

14 What changes might occur in a rural area experiencing **population decline**?   [4]

15 Giving examples, suggest some of the **social** and **economic** benefits from an improvement in the rail infrastructure in the UK.   [4]

Total Marks _____ / 47

The Challenge of Resource Management

1 Define the term 'carbon footprint'. [2]

2 Suggest why some of the food eaten in the UK has a high carbon footprint. [3]

3 Suggest reasons for the increase in the **domestic** demand for water in the UK. [2]

4 Explain the difficulties of matching **supply and demand** for water in the UK. [6]

5 Suggest why the UK's reliance on **fossil fuels** is seen as a security risk. [4]

6 Study the figure below. Suggest reasons why some regions of the world have **water stress** or **scarcity**. [4]

Fresh Water in the World
Access to renewable water sources (m³ per person, per year)

- No data
- 0 – 1000 (Scarcity)
- 1000 – 2500 (Stress / vulnerability)
- 2500 – 15 000
- 15 000 – 50 000+

7 What are **renewables**? [3]

8 Outline why some people have objections to **fracking**. [3]

Video Solution   Question 12

9 Many countries in Africa suffer from **food insecurity**. Suggest **two** reasons for this.　　　[4]

10 Describe what conditions might lead to a **famine**.　　　[4]

11 One way of increasing food supply is by use of irrigation. What is **irrigation**?　　　[1]

12 Suggest how **biotechnology** can help to increase food supply.　　　[3]

13 Using an example you have studied, show how **urban farming** is important in feeding those who are less well off.　　　[3]

14 Water usage is increasing worldwide but some places suffer from water scarcity. Using examples, suggest what can be done to ensure **sustainable water supplies**. [6]

15 Suggest reasons for the increase in **energy consumption** worldwide. [3]

16 Describe how energy usage can be made more **sustainable**. [5]

Total Marks _____ / 56

1 For your geographical enquiry, describe and explain patterns in your data and any anomalies which did not correspond to the main patterns. [8]

Total Marks _____ / 8

Collins

GCSE
GEOGRAPHY
Paper 1 Living with the physical environment

Time allowed: 1 hour 30 minutes

Materials

For this paper you must have:

- a pencil
- a rubber
- a ruler.

You may use a calculator.

Instructions

- Use black ink or a black ball-point pen.
- Answer **all** questions in Section A and Section B.
- Answer **two** questions in Section C.
- Cross through any work you do not want to be marked.

Information

- The marks for questions are shown in brackets.
- The total number of marks available for this paper is 88.
- Spelling, punctuation, grammar and specialist terminology will be assessed in Question 01.10.

Advice

- For the multiple-choice questions, completely fill in the circle alongside the appropriate answer(s).

CORRECT METHOD ●   WRONG METHODS ⊗ ⊙ ☰ ✓

- If you want to change your answer, you must cross out your original answer as shown. ⊠

- If you wish to return to an answer previously crossed out, ring the answer you now wish to select as shown. ⊘

Name:

Section A: The challenge of natural hazards

Answer **all** questions in this section.

Question 1   The challenge of natural hazards

Figure 1 shows some of the world's tectonic plates and their direction of movement. Two countries which experienced earthquakes in 2018, the UK and Japan, are indicated.

Figure 1

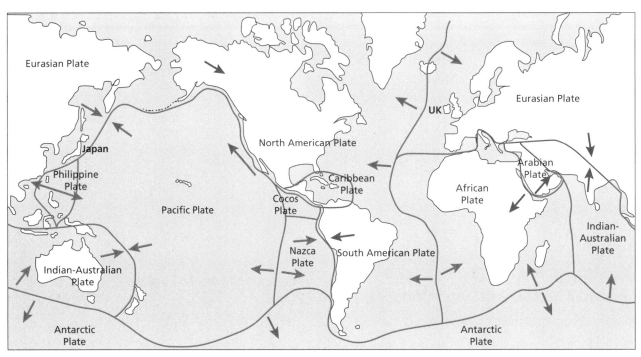

0 1 · 1   Using **Figure 1**, which one of the following statements is true?   [1 mark]

Shade **one** circle only.

A   There is a constructive plate margin between the Pacific Plate and the Philippine Plate.   ○

B   There is a constructive plate margin between the Pacific Plate and the North American Plate.   ○

C   There is a constructive plate margin between the Eurasian Plate and the North American Plate.   ○

D   There is a constructive plate margin between the Arabian Plate and the Eurasian Plate.   ○

0 1 · 2 Using **Figure 1**, name the type of plate margin between the South American and the Nazca plates. **[1 mark]**

0 1 · 3 With the help of **Figure 1**, explain why Japan experiences more earthquakes than the UK. **[2 marks]**

Study **Figure 2**, a photograph showing the aftermath of an earthquake in Sapporo, Japan, in September 2018.

Figure 2

Question 1 continues on the next page

0 1 · 4 Suggest how tectonic hazards can have primary and secondary effects.

Use **Figure 2** and your own understanding.   **[6 marks]**

_____

_____

_____

_____

_____

_____

_____

_____

_____

_____

Study **Figure 3**, a map showing global air circulation and surface winds.

Figure 3

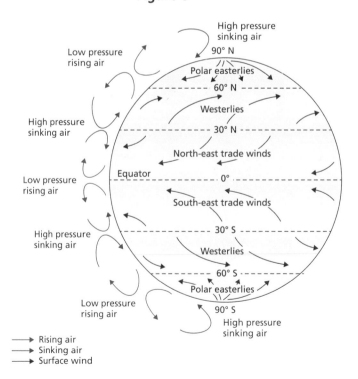

`0 1 · 5` Using **Figure 3**, which **two** of the following statements are true? [2 marks]

Shade **two** circles only.

A   Air sinks at the Equator.   ◯

B   There is high pressure at the North Pole.   ◯

C   There is sinking air in areas of low pressure.   ◯

D   The north-east trade winds blow from 30°N towards the Equator. ◯

E   Westerlies blow from the Equator towards 30° North and South. ◯

F   There is low pressure at 30° North and South.   ◯

`0 1 · 6` Using **Figure 3**, describe the link between air pressure and surface winds. [1 mark]

....

....

`0 1 · 7` Explain the causes of tropical storms.   [4 marks]

....

....

....

....

....

....

....

....

Question 1 continues on the next page

Study **Figure 4**, a table listing some Atlantic Ocean tropical storms since 2004.

Figure 4

Tropical storm	Number of deaths	Maximum wind speed (miles per hour)
Hurricane Charley, 2004	15	150
Hurricane Katrina, 2005	1200	175
Hurricane Ike, 2008	195	145
Hurricane Igor, 2010	4	155
Hurricane Irene, 2011	49	121
Superstorm Sandy, 2012	285	110
Hurricane Matthew, 2016	603	165
Hurricane Michael, 2018	43	161

0 1 · 8 'As the maximum wind speed of tropical storms increases, so do the number of deaths.'

Do you agree with this statement?

Use evidence from **Figure 4** to support your answer. **[2 marks]**

0 1 · 9 Suggest how climate change might affect the distribution and intensity of tropical storms. **[2 marks]**

Study **Figure 5**, photographs showing strategies used to manage climate change.

Figure 5

Mitigation strategies

Alternative energy production

Planting trees

Adaptation strategies

Reducing risks from rising sea levels

Change in agricultural systems

Question 1 continues on the next page

0 1 · 10 'Managing climate change involves both mitigation (reducing causes) and adaptation (responding to change).'

Do you agree with this statement?

Explain your answer.

Use **Figure 5** and your own understanding.   **[9 marks] [+ 3 SPaG marks]**

End of Section A

Section B: The living world

Answer **all** questions in this section.

Question 2 The living world

Study **Figure 6**, a world map showing the distribution of two types of forest biome.

Figure 6

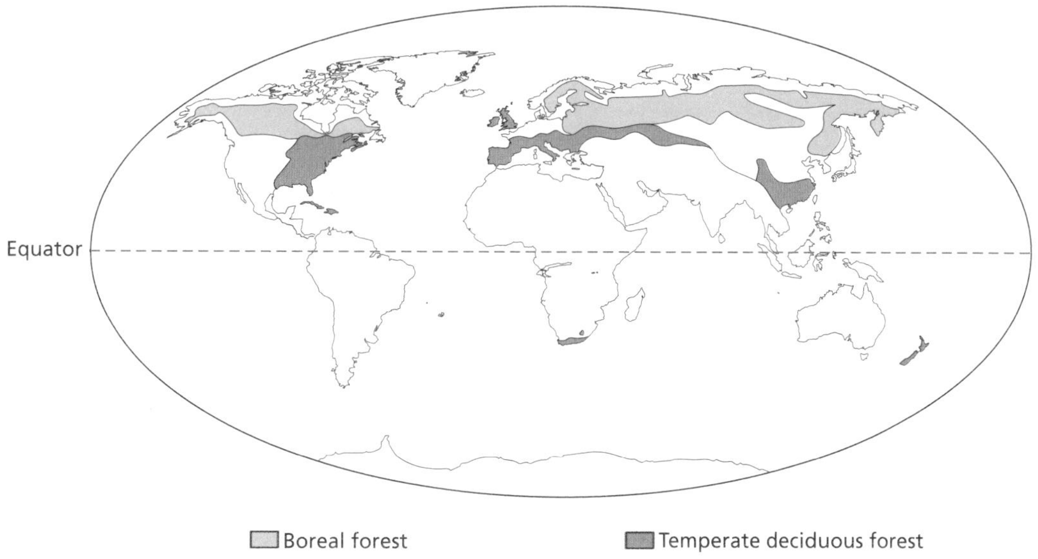

Boreal forest   Temperate deciduous forest

---

0 2 · 1 Using **Figure 6**, which of the statements below is **true**?   [1 mark]

Shade **one** circle only.

A   There is temperate deciduous forest in every continent.   ◯

B   There is no boreal forest in the Southern Hemisphere.   ◯

C   Worldwide, there is more temperate deciduous forest than boreal forest.   ◯

D   Boreal forest is always found on the western side of continents.   ◯

0 2 · 2 Using **Figure 6**, which of the statements below about the climate of temperate deciduous forest is **true**?   [1 mark]

Shade the circle next to the correct statement.

A   Short mild summers, long cold winters, often below 0°C.   ◯

B   Warm summers, mild winters, rain falling all year round.   ◯

Question 2 continues on the next page

| 0 | 2 | · | 3 | What is an ecosystem?   [1 mark]

| 0 | 2 | · | 4 | State **one** role of decomposers in an ecosystem.   [1 mark]

Study **Figure 7**, which shows part of the food web for an area of Scottish upland.

Figure 7

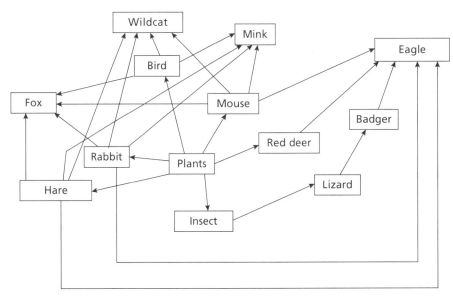

| 0 | 2 | · | 5 | Using **Figure 7**, give **one** example of a primary consumer.   [1 mark]

| 0 | 2 | · | 6 | Suggest what would happen in the food web shown in **Figure** 7 if red deer became extinct.   [2 marks]

Study **Figure 8**, which shows a climate graph for Manaus, a city in the Amazon Rainforest in Brazil.

Figure 8

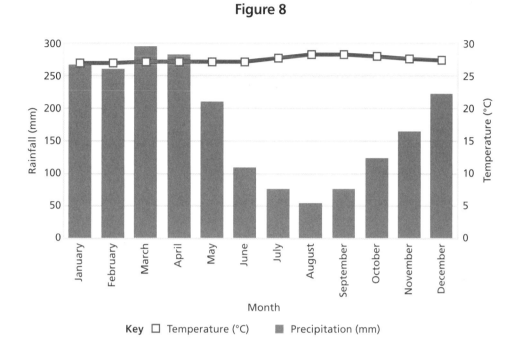

Key ☐ Temperature (°C)   ■ Precipitation (mm)

0 2 · 7 Using **Figure 8**, describe the pattern of rainfall.   **[2 marks]**

0 2 · 8 Give **one** reason why tropical rainforests have high temperatures throughout the year.   **[1 mark]**

Question 2 continues on the next page

Study **Figure 9**, which shows photographs of the Amazon Rainforest in Brazil.

Figure 9

0 2 · 9 Using **Figure 9** and your own understanding, explain how deforestation has economic and environmental impacts. **[6 marks]**

0 2 · 10 'Plants and animals adapt to survive in harsh physical conditions.'

Choose **one** of the following environments.

Hot desert environment ☐

Cold environment ☐

Tick the box to show which environment you have chosen.

Using a case study, to what extent do you agree with this statement? **[9 marks]**

_____

_____

_____

_____

_____

_____

_____

_____

_____

_____

_____

_____

_____

_____

_____

_____

_____

End of Section B

Section C: Physical landscapes in the UK

Answer **two** questions from the following:
Question 3 (Coasts), Question 4 (Rivers), Question 5 (Glacial)

Shade the circle below to indicate which **two** optional questions you will answer.

Question 3 ⬭   Question 4 ⬭   Question 5 ⬭

Question 3 Coastal landscapes in the UK

Study **Figure 10**, a 1: 25 000 Ordnance Survey map extract of an area near to Scarborough in North Yorkshire.

Figure 10

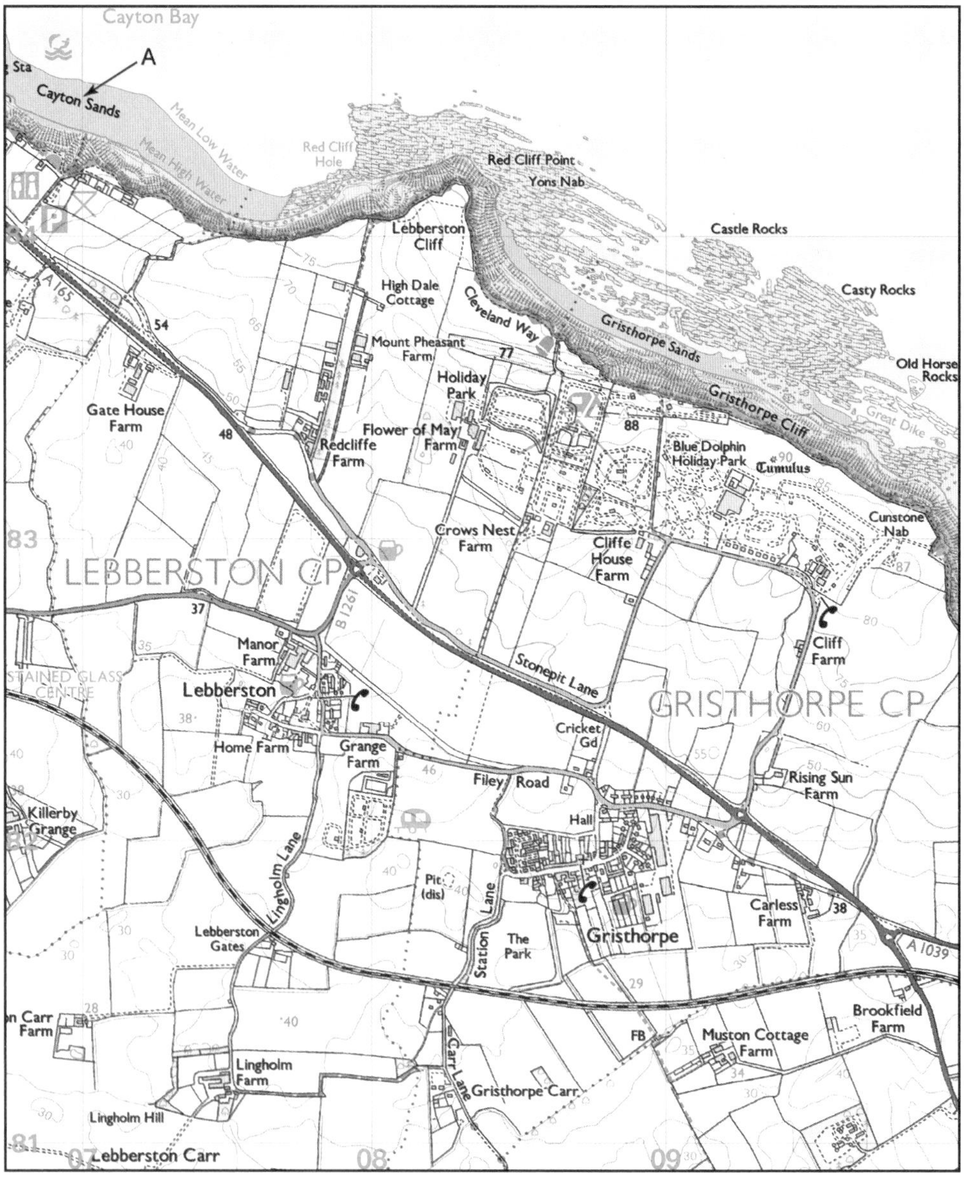

Scale 1 : 25 000
4 centimetres to 1 kilometre (one grid square)

0 3 · 1 Using **Figure 10**, give the six-figure grid reference for Red Cliff Hole. **[1 mark]**

Shade **one** circle only.

A   073844   ○   **B**   085837   ○

C   079842   ○   **D**   099839   ○

0 3 · 2 Using **Figure 10**, which of the following coastal features is **not** shown in grid square 0983?   **[1 mark]**

Shade **one** circle only.

A   A beach   ○   **B**   A wave cut platform   ○

C   A cliff   ○   **D**   A spit   ○

0 3 · 3 Using **Figure 10**, what is the distance (in metres) between the mean high water and the mean low water at **point A** in grid square 0784?   **[1 mark]**

0 3 · 4 Using **Figure 10**, describe **one** piece of evidence which suggests that the area on the map attracts tourists.   **[2 marks]**

0 3 · 5 Explain the formation of sand dunes.   **[4 marks]**

Question 3 continues on the next page

Study **Figure 11**, photographs showing some hard engineering strategies.

Figure 11

0 3 · 6 Discuss the costs and benefits of hard engineering strategies in protecting coastlines.

[6 marks]

Question 4   River landscapes in the UK

Study **Figure 12**, a 1: 25 000 Ordnance Survey map extract of part of the south Pennines in Lancashire.

Figure 12

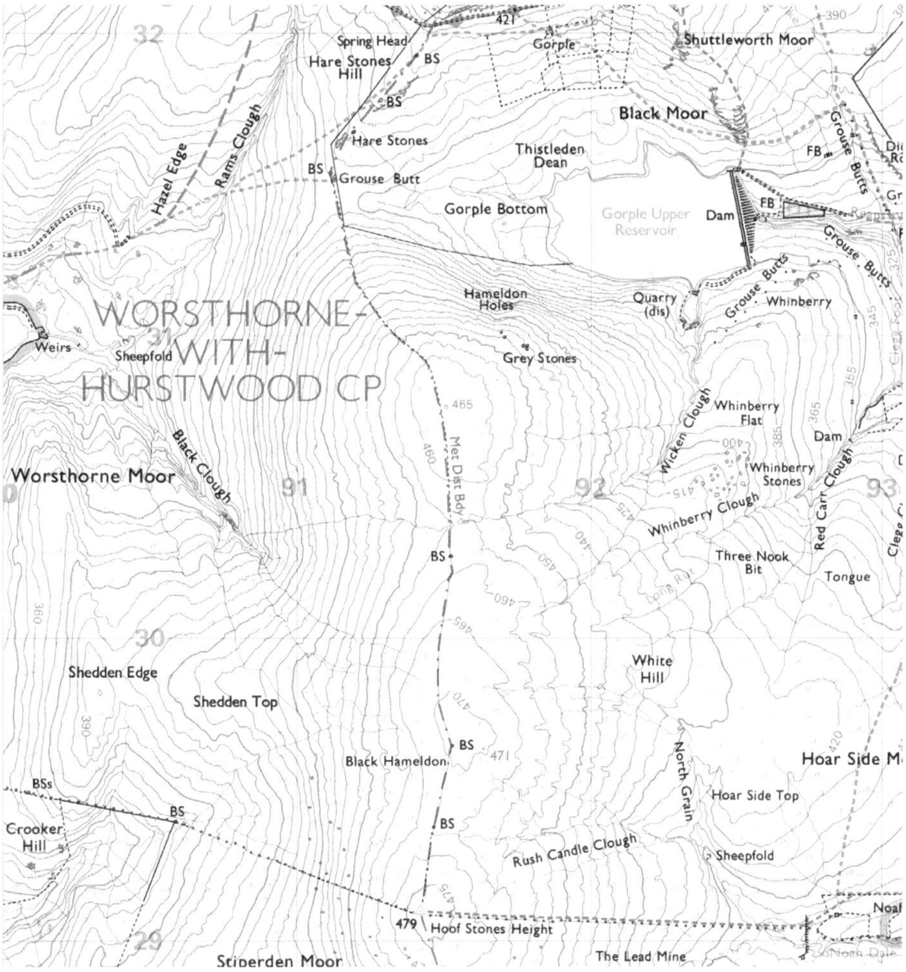

Scale 1 : 25 000
4 centimetres to 1 kilometre (one grid square)

0 4 · 1 In grid square 9131 of **Figure 12**, which area has the steepest slopes? **[1 mark]**

Shade **one** circle only.

A   South-East   ⬭

B   North-East   ⬭

C   South-West   ⬭

D   North-West   ⬭

Question 4 continues on the next page

0 4 · 2 Using **Figure 12**, describe the shape of the land around Black Clough river in grid square 9030. **[1 mark]**

0 4 · 3 Using **Figure 12**, which location is at a confluence? **[1 mark]**

Shade **one** circle only.

A   914293   ○

B   905306   ○

C   920305   ○

D   929311   ○

0 4 · 4 Using **Figure 12** and your own understanding, describe the river erosion processes which will be taking place in this area. **[2 marks]**

0 4 · 5 Explain how river levées are formed. **[4 marks]**

Study **Figure 13**, photographs showing some flood management strategies.

Figure 13

0 4 · 6 Discuss how flood management schemes can lead to social, economic and environmental issues.

Use **Figure 13** and an example you have studied. [6 marks]

Section C continues on the next page

Question 5 Glacial landscapes in the UK

Study **Figure 14**, a 1: 25 000 Ordnance Survey map extract of part of Snowdonia in north Wales.

Figure 14

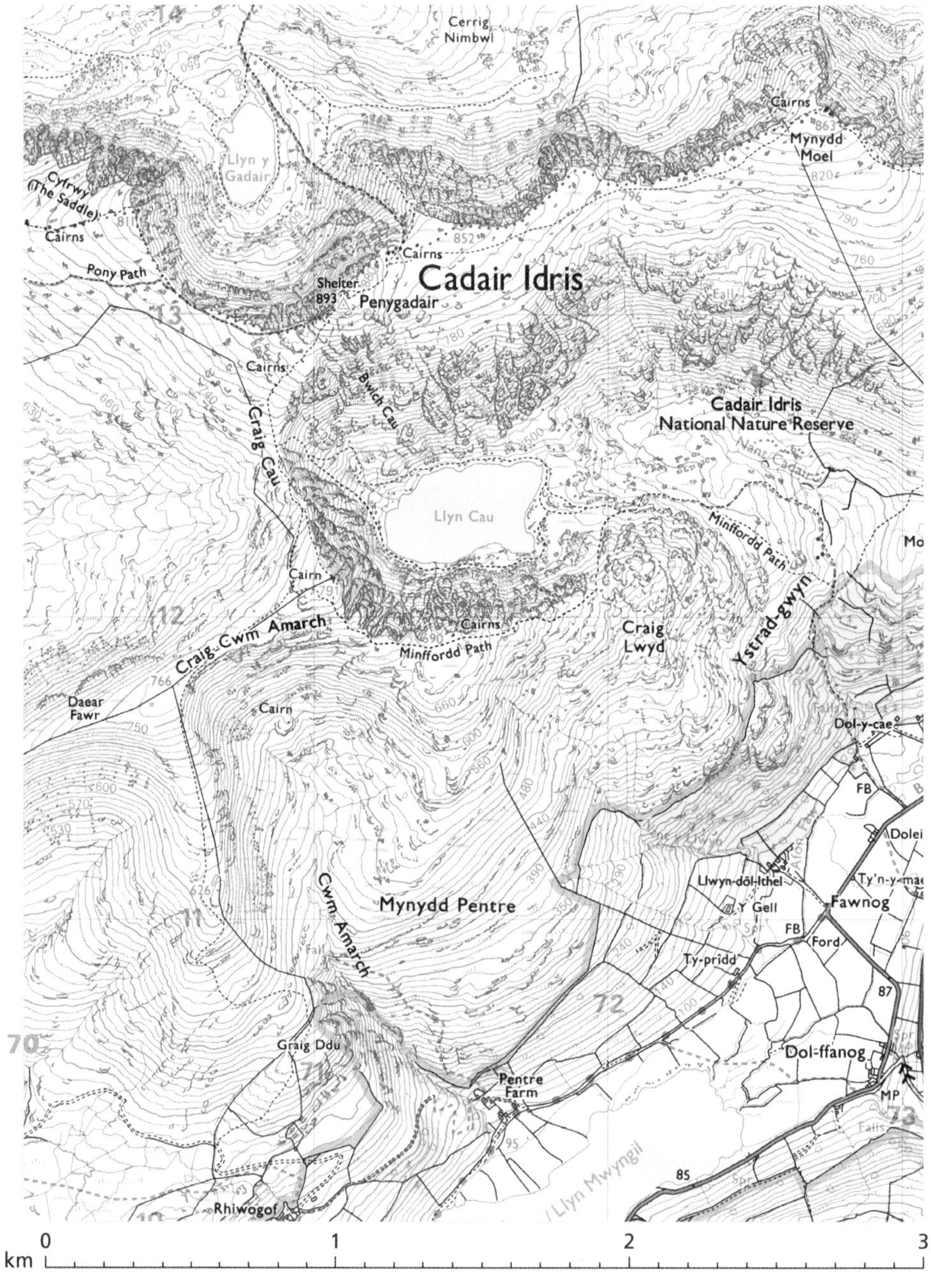

Scale 1 : 25000
4 centimetres to 1 kilometre (one grid square)

0 5 · 1 Using **Figure 14**, which grid reference is at the centre of a corrie lake? **[1 mark]**

Shade **one** circle only.

A 707135 ⬭ B 705132 ⬭

C 711125 ⬭ D 716113 ⬭

0 5 . 2 Using **Figure 14**, what is the difference in height between the spot height at 711131 and the spot height at 705118? **[1 mark]**

0 5 . 3 Using **Figure 14**, describe the land uses shown in grid square 7210. **[2 marks]**

0 5 . 4 Which of these is a process of glacial erosion? **[1 mark]**

Shade **one** circle only.

A   Abrasion   ○

B   Freeze-thaw   ○

C   Rotational slip   ○

D   Deposition   ○

0 5 . 5 Explain the formation of a drumlin. **[4 marks]**

Question 5 continues on the next page

Practice Exam Paper 1

Study **Figure 15**, photographs showing some tourism management strategies.

Figure 15

0 5 · 6 Discuss the strategies used to manage the impact of tourism in glaciated upland areas.

Use **Figure 15** and an example you have studied. **[6 marks]**

....

....

....

....

....

....

....

....

....

....

End of Section C

END OF QUESTIONS

Collins

GCSE
GEOGRAPHY

Paper 2 Challenges in the human environment

Time allowed: 1 hour 30 minutes

Materials

For this paper you must have:

- a pencil
- a rubber
- a ruler.

You may use a calculator.

Instructions

- Use black ink or a black ball-point pen.
- Answer **all** questions in Section A and Section B.
- Answer Question 3 and **one other** question in Section C.
- Cross through any work you do not want to be marked.

Information

- The marks for questions are shown in brackets.
- The total number of marks available for this paper is 88.
- Spelling, punctuation, grammar and specialist terminology will be assessed in Question 01.10.

Advice

- For the multiple-choice questions, completely fill in the circle alongside the appropriate answer(s).

CORRECT METHOD   WRONG METHODS

- If you want to change your answer, you must cross out your original answer as shown.

- If you wish to return to an answer previously crossed out, ring the answer you now wish to select as shown.

Name:

Section A: Urban issues and challenges

Answer **all** questions in this section.

Question 1 Urban issues and challenges

Study **Figure 1**, a map showing global natural population increase.

Figure 1

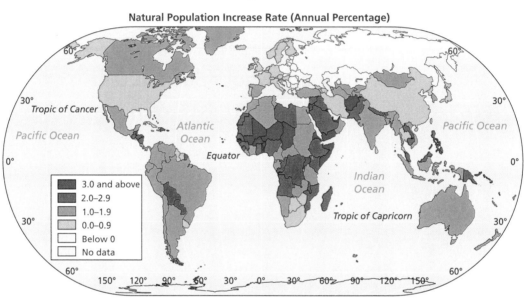

Natural Population Increase Rate (Annual Percentage)

0 1 · 1 Using **Figure 1**, in which of the continents shown is natural increase highest? **[1 mark]**

Shade **one** circle only.

A Europe ⬭

B South America ⬭

C Africa ⬭

D North America ⬭

E Asia ⬭

0 1 · 2 Using **Figure 1**, what is the UK's natural population increase? **[1 mark]**

0 1 · 3 Outline **one** reason why rates of natural increase are high in many cities in lower income countries (LICs) and newly emerging economies (NEEs). **[2 marks]**

Study **Figure 2**, a photograph of a squatter settlement in Rio de Janeiro, Brazil.

Figure 2

0 1 · 4 Using **Figure** 2, suggest **one** problem faced by people in Rio de Janeiro as a result of urban growth. **[2 marks]**

_____

_____

_____

0 1 · 5 Using an example, describe how urban planning is improving the quality of life for the urban poor. **[6 marks]**

_____

_____

_____

_____

_____

_____

_____

_____

Question 1 continues on the next page

0 1 · 6 What is 'urban greening'? [1 mark]

Study **Figure 3**, a photograph of One Central Park in Sydney, Australia.

Figure 3

0 1 · 7 Explain why creating green space is important for sustainable urban living.

Use **Figure 3** and your own understanding. [4 marks]

0 1 · 8 Complete the following fact file for a UK city that you have studied. **[2 marks]**

Name of UK city	
Location in the UK	
Importance in the UK	

0 1 · 9 Describe the impacts of international migration on the character of a UK city that you have studied. **[2 marks]**

....

....

....

....

Study **Figure 4**, photographs of Ancoats in Manchester before and after regeneration.

Figure 4

Question 1 continues on the next page

0 1 · 10 Assess the extent to which a regeneration project can solve urban problems.

Use **Figure 4** and an example you have studied.   **[9 marks] [+ 3 SPaG marks]**

End of Section A

Section B: The changing economic world

Answer **all** questions in this section.

Question 2   The changing economic world

Study **Figure 5**, a table showing Gross National Income (GNI) data for selected countries in 2020.

Figure 5

Name of Country	GNI (US$ per person)
Argentina	9070
Bangladesh	2030
Denmark	63,010
Ethiopia	890
India	1920
Italy	32,290
Kenya	1840
Mali	830
New Zealand	41,550
Portugal	21,790
Saudi Arabia	21,930

`0 2 · 1` Calculate the median value for the GNI data in **Figure 5**.   [2 marks]

Median = _____

`0 2 · 2` Give **two** social measures of development.   [2 marks]

_____

_____

Question 2 continues on the next page

0 2 · 3 Outline the limitations of economic measures of development. **[3 marks]**

0 2 · 4 Explain how fair trade can reduce the development gap. **[4 marks]**

0 2 · 5 Describe the environmental context of a named lower income country (LIC) or newly emerging economy (NEE).

Name of country: .... **[2 marks]**

0 2 · 6 Using a case study of a LIC or NEE, discuss the advantages and disadvantages of transnational corporations to the country. **[6 marks]**

Study **Figure 6**, pie charts showing how the UK's employment structure has changed over time.

Figure 6

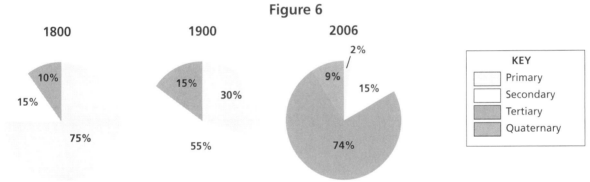

0 2 · 7 Using **Figure 6**, which **two** of the following statements are true? **[2 marks]**

Shade **two** circles only.

A   In 1800, most people were employed in the secondary sector.   ○

B   The largest employment sector in 2006 was the tertiary sector.   ○

C   Employment in the primary sector has decreased over time.   ○

D   In 1900, most people were employed in the quaternary sector.   ○

E   Employment in the primary sector has stayed constant over time.   ○

F   The tertiary sector has always employed the most people.   ○

Question 2 continues on the next page

0 2 · 8 To what extent have strategies to reduce the UK's north-south divide been successful? **[9 marks]**

End of Section B

Section C: The challenge of resource management

Answer Question 3 and **either** Question 4 **or** Question 5 **or** Question 6.

Question 3   The challenge of resource management

Study **Figure 7**, a graph showing the daily caloric intake of selected countries around the world.

Figure 7

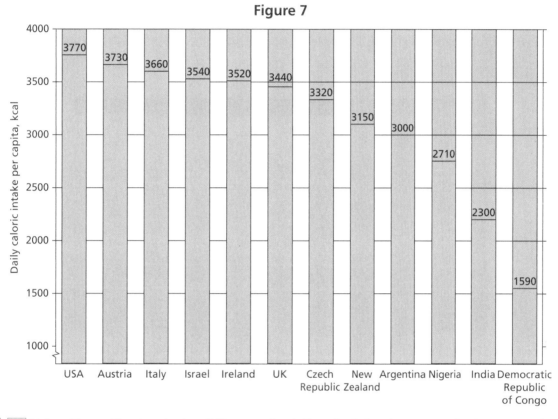

| 0 | 3 |·| 1 | Using **Figure 7**, what is the difference in daily caloric intake per capita between the USA and the Democratic Republic of Congo? **[1 mark]**

| 0 | 3 |·| 2 | Using **Figure 7** and your own understanding, suggest how daily caloric intake can influence well-being. **[3 marks]**

Question 3 continues on the next page

0 3 · 3 Outline **one** disadvantage of importing food from other countries. **[2 marks]**

Study **Figure 8**, a map showing levels of water stress across England and parts of Wales.

Figure 8

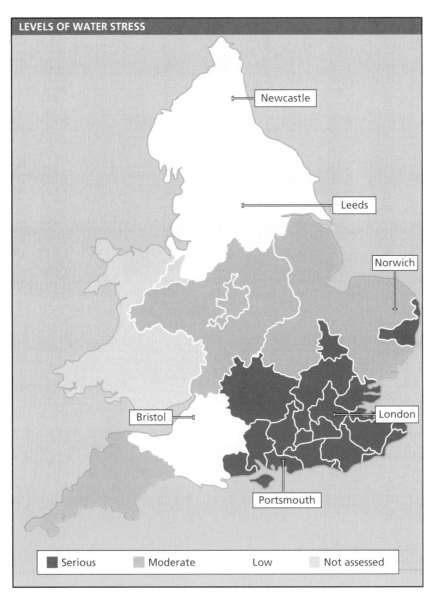

0 3 · 4 Using **Figure 8** and your own understanding, discuss the issues arising from the UK's changing demand for water. **[6 marks]**

_____

_____

_____

_____

_____

_____

_____

_____

_____

Section C continues on the next page

Practice Exam Paper 2

Answer **either** Question 4 (Food) **or** Question 5 (Water) **or** Question 6 (Energy).

Question 4 Food

Study **Figure 9**, a map showing the percentage of the population that are undernourished in the continent of Africa.

Figure 9

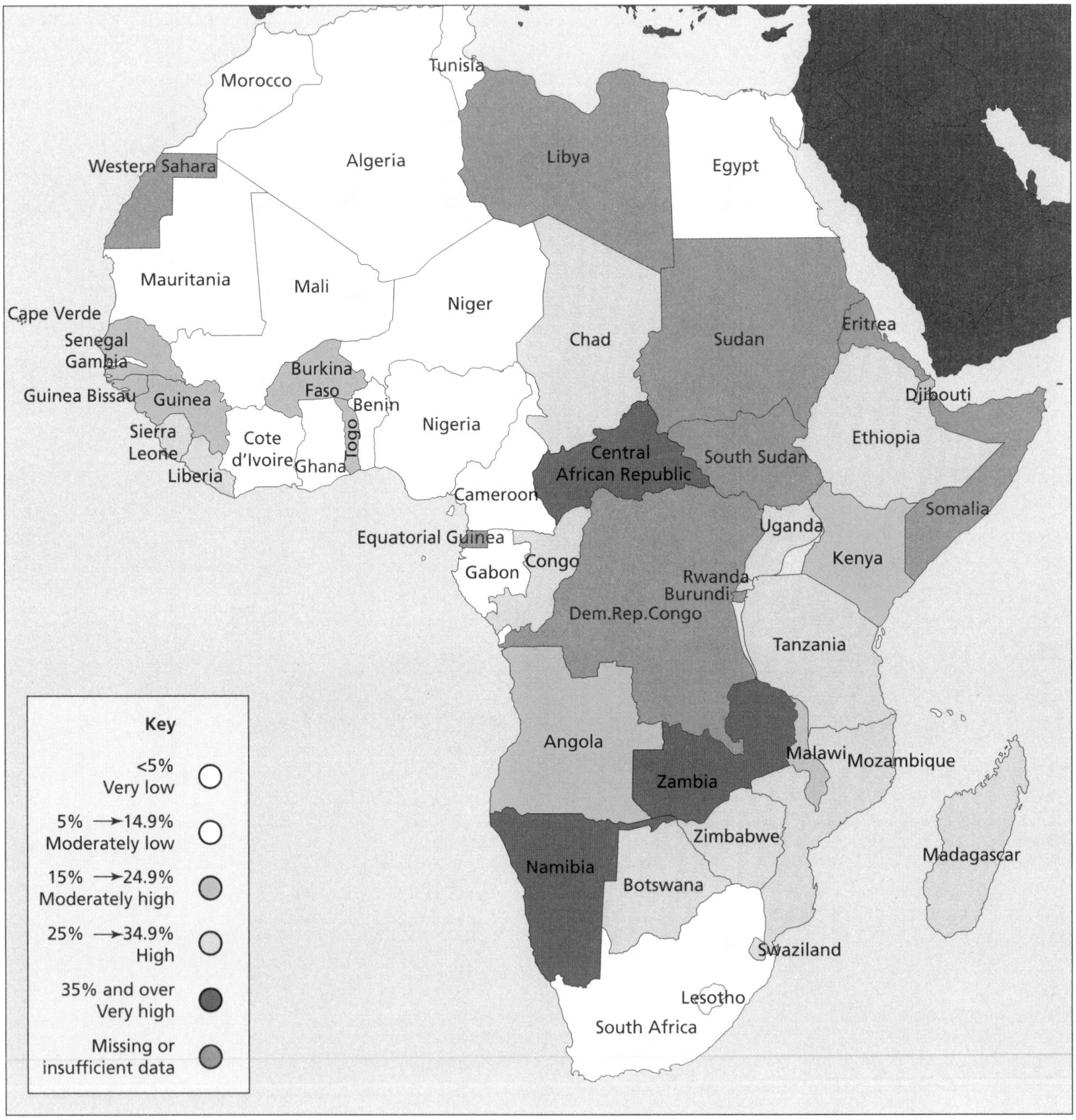

0 4 · 1 Using **Figure 9**, name **one** country where less than 5% of the population is undernourished. **[1 mark]**

....

0 4 · 2 Using **Figure 9**, how many of the African countries are shown to have over 35% of their population undernourished? [1 mark]

Shade **one** circle only.

A   3   ⃝

B   5   ⃝

C   7   ⃝

D   8   ⃝

0 4 · 3 Using **Figure 9**, describe the distribution of the countries which had less than 5% of their population undernourished. [2 marks]

....

....

....

....

0 4 · 4 Suggest **one** reason for differences in undernourishment between countries. [2 marks]

....

....

....

....

0 4 · 5 What is meant by 'food insecurity'? [1 mark]

....

....

Question 4 continues on the next page

Study **Figure 10a** and **Figure 10b**, photographs showing strategies to increase food supply.

| Figure 10a – Using hydroponics to grow crops | Figure 10b – Irrigating crops in Burkina Faso |

0 4 · 6 Using **Figure 10a** and **Figure 10b**, explain how countries can increase their food supply.

[6 marks]

Question 5   Water

Study **Figure 11**, a map showing predicted levels of water stress across Europe in 2040.

Figure 11

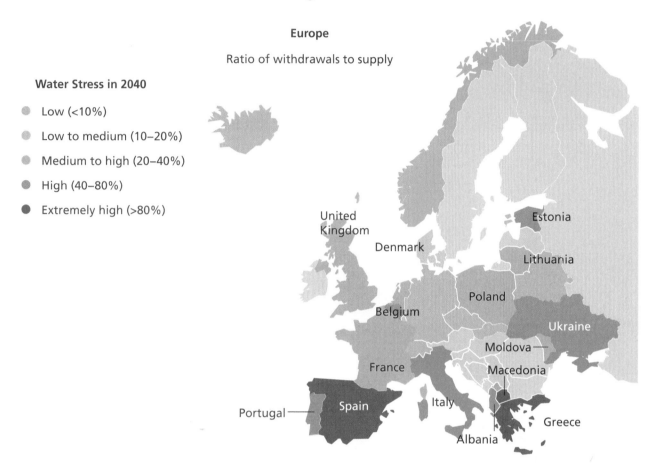

0 5 · **1** Using **Figure 11**, name **one** country with a water stress rating of over 80%.   **[1 mark]**

0 5 · **2** Using **Figure 11**, how many European countries have a water stress ratio of less than 10%?   **[1 mark]**

Shade **one** circle only.

A   5

B   7

C   10

D   12

Question 5 continues on the next page

Practice Exam Paper 2

0 5 . 3 Using **Figure 11**, describe the distribution of the countries which have a water stress ratio of less than 10%.

[2 marks]

0 5 . 4 Suggest **one** reason for differences in water stress between countries.

[2 marks]

0 5 . 5 What is meant by 'water insecurity'?

[1 mark]

Study **Figure 12a** and **Figure 12b**, photographs showing strategies to increase water supply.

Figure 12a – A Desalination Plant at Arrecife, Lanzarote, Canary Islands	Figure 12b – Ridgegate Reservoir near Macclesfield in Cheshire, UK

0 5 · 6 Using **Figure 12a** and **Figure 12b**, explain how higher income countries can increase their water supply. **[6 marks]**

Question 6   Energy

Study **Figure 13**, a map showing projected global change in primary energy demand.

Figure 13

Change in primary energy demand, 2016–40 (Million tonnes of oil equivalent)

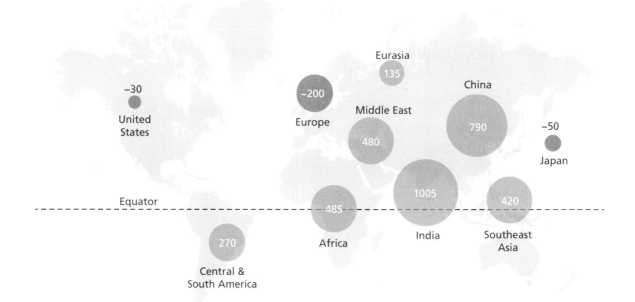

0 6 · 1 Using **Figure 13**, which part of the world is projected to experience the highest increase in energy demand? **[1 mark]**

0 6 · 2 Using **Figure 13**, how many parts of the world are predicted to see a fall in energy demand? **[1 mark]**

Shade **one** circle only.

A   2   ◯

B   3   ◯

C   4   ◯

D   5   ◯

0 6 · 3 Using **Figure 13**, describe the distribution of the parts of the world which are predicted to see a fall in energy demand. **[2 marks]**

....

....

....

....

0 6 · 4 Suggest **one** reason for differences in predicted energy demand changes between different countries. **[2 marks]**

....

....

....

....

0 6 · 5 What is meant by 'energy insecurity'? **[1 mark]**

....

....

Question 6 continues on the next page

Study **Figure 14a** and **Figure 14b**, photographs showing strategies to increase energy supply.

Figure 14a – North Hoyle Wind Farm off the coast of North Wales	Figure 14b – A nuclear power plant near Thionville in France

[0][6] · [6] Using **Figure 14a** and **Figure 14b**, explain how countries can increase their energy supply. **[6 marks]**

End of Section C

END OF QUESTIONS

Collins

GCSE
GEOGRAPHY
Paper 3 Geographical applications

Time allowed: 1 hour 15 minutes

Materials

For this paper you must have:

- a clean copy of the pre-release resources booklet (see page 225).
- a pencil
- a rubber
- a ruler.

You may use a calculator.

Instructions

- Use black ink or a black ball-point pen.
- Answer **all** questions.
- Cross through any work you do not want to be marked.

Information

- The marks for questions are shown in brackets.
- The total number of marks available for this paper is 76.
- Spelling, punctuation, grammar and specialist terminology will be assessed in Questions 03.3 and 05.4

Advice

- For the multiple-choice questions, completely fill in the circle alongside the appropriate answer(s).

CORRECT METHOD   WRONG METHODS

- If you want to change your answer, you must cross out your original answer as shown.

- If you wish to return to an answer previously crossed out, ring the answer you now wish to select as shown.

Name:

Practice Exam Paper 3

Section A: Issue evaluation

Answer **all** questions in this section.

Study **Figure 1**, a graph showing the sources of energy used to create electricity in the UK between 1996 and 2017.

Figure 1

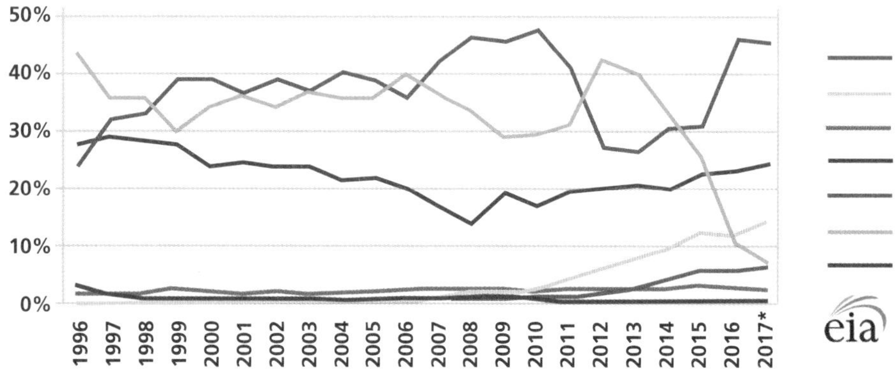

50%
40%
30%
20%
10%
0%
1996
1997
1998
1999
2000
2001
2002
2003
2004
2005
2006
2007
2008
2009
2010
2011
2012
2013
2014
2015
2016
2017*
Biomass
Wind/solar
Hydro
Nuclear
Natural gas
Coal
Oil and other
eia

*note 2017 values are estimates based on data through September

Source: U.S. Energy Information Administration based on Digest of UK Energy Statistics and National Statistics: Energy Trends

0 1 · 1 Using **Figure 1**, in which year did coal and nuclear provide the same amount of electricity? **[1 mark]**

Shade **one** circle only.

A	2015	○	**B**	2017	○
C	2000	○	**D**	2008	○

0 1 · 2 What is meant by 'fossil fuel'? **[1 mark]**

....

....

Study **Figure 2**, a chart which shows the origin of gas supplies in the UK in 2016.

Figure 2

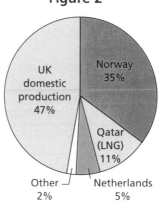

UK domestic production 47%
Norway 35%
Qatar (LNG) 11%
Other 2%
Netherlands 5%

0 1 · 3 Using **Figure 2**, describe the pattern shown on the chart. **[2 marks]**

0 1 · 4 Outline **one** disadvantage of importing energy from other countries. **[2 marks]**

0 1 · 5 'Economic and environmental issues are caused by the exploitation of energy sources.'

Discuss this statement. **[6 marks]**

Section A continues on the next page

Practice Exam Paper 3

0 2 · 1 What is meant by 'renewable energy'? [2 marks]

0 2 · 2 Suggest reasons for the growing significance of renewable energy sources in the UK. [6 marks]

Study **Figure 3**, a table showing the changes in UK wind power capacity, generation and percentage of total electricity provided by wind power.

Figure 3

Year	Capacity (MW)	Generation (GW/h)	% of total electricity use
2011	6,540	12,675	3.81
2012	8,871	20,710	5.52
2013	10,976	24,500	7.39
2014	12,440	28,100	9.30
2015	13,602	40,442	11.0
2016	16,218	37,368	12.0
2017	19,837	49,607	17.0
2018	21,606	56,907	17.1
2019	23,882	63,795	19.7
2020	24,485	75,369	24.1

0 3 · 1 Using **Figure 3**, what was the increase in wind power as a percentage of total electricity use between 2011 and 2020? **[1 mark]**

Shade **one** circle only.

A   24.1   ⬭

B   3.81   ⬭

C   27.91   ⬭

D   20.29   ⬭

Study **Figure 4**, a photograph of a wind farm in the Scottish Highlands.

Figure 4

0 3 · 2 Using **Figure 4** and your own understanding, suggest reasons why the UK has got good potential for developing wind energy. **[4 marks]**

....

....

....

....

....

....

....

....

Section A continues on the next page

0 3 . 3 'Developing wind energy projects in the Scottish Highlands are a good idea.'

Do you agree with this statement? **[9 marks] [+ 3 SPaG marks]**

Yes ☐   No ☐

Tick the box to show your choice.

Use evidence from the resources booklet and your own understanding to explain your answer.

End of Section A

Section B: Fieldwork
Answer **all** questions in this section.

A group of students visit part of Snowdonia in North Wales to carry out a fieldwork enquiry. They want to investigate the hypothesis that 'the size of pebbles decreases in size as distance down the valley increases'.

In order to do this, the students measure the long axis (longest side) of stones in material deposited by glaciers. They take the measurement of 100 stones from each of five evenly spaced locations shown in **Figure 5** below.

Figure 5

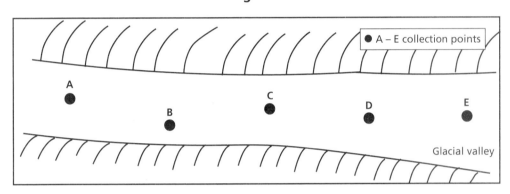

0 4 · 1 Identify the type of sampling method used in **Figure 5**.   [1 mark]

Shade **one** circle only.

A	Random sampling	◯
B	Stratified sampling	◯
C	Systematic sampling	◯
D	Opportunity sampling	◯

0 4 · 2 Suggest why the type of sampling shown in **Figure 5** is not always possible in a fieldwork enquiry.   [2 marks]

....

....

....

....

Section B continues on the next page

Figure 6 shows the mean length of stones, in each of the five locations.

Figure 6

	Location A	Location B	Location C	Location D	Location E
Distance down the valley	500 m	1000 m	1500 m	2000 m	2500 m
Mean	72 mm	94 mm	103 mm	166 mm	208 mm

0 4 · 3 Suggest a suitable method for presenting the data shown in **Figure 6**.

Give a reason for your choice. **[2 marks]**

0 4 · 4 Suggest **one** advantage and **one** disadvantage of using the mean as a measure of central tendency. **[2 marks]**

Advantage

Disadvantage

0 4 · 5 Outline the conclusions that the students could make from the data in **Figure 6**. **[2 marks]**

On a different day, the students wanted to investigate the quality of two footpaths. Students asked 50 people their opinion of the quality of two footpaths.

Study **Figure 7**, a table showing the results of their survey.

Figure 7

Quality of footpath	Footpath A	Footpath B
Very good	25	10
Good	15	18
Fair	8	12
Poor	2	10

0 4 · 6 Complete **Figure 8** below to show the survey results for Footpath B. [1 mark]

Figure 8

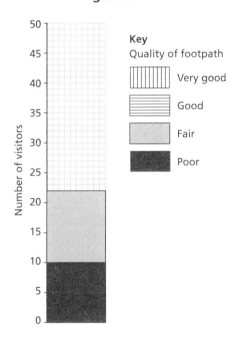

0 4 · 7 What percentage of visitors thought the quality of Footpath B was good or very good? [1 mark]

0 4 · 8 Suggest **one** advantage of using a bar chart to present the data shown in **Figure 7**. [1 mark]

Section B continues on the next page

Figure 9 shows the change in width along the first 1000 metres of footpaths A and B. The footpath routes have been made into straight lines to make it easier to interpret the results.

Figure 9

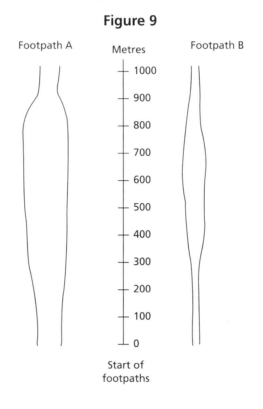

0 4 . 9 Using **Figure 9**, compare the width of the two footpaths. **[4 marks]**

Write the title of your physical geography fieldwork enquiry.

Title of physical fieldwork enquiry: ....

0 5 . 1 State **one** potential risk you identified in your physical geography fieldwork and **explain** the measure(s) you used to reduce it. **[3 marks]**

Risk

Measure(s) used to reduce risk

0 5 . 2 Assess the effectiveness of your data collection method(s). **[6 marks]**

Section B continues on the next page

Write the title of your human geography fieldwork enquiry.

Title of human fieldwork enquiry: ....

....

0 5 . 3 Explain why the chosen location was suitable for the collection of data.   **[2 marks]**

0 5 . 4 For **one** of your fieldwork enquiries, to what extent did the data collected allow you to reach valid conclusions?   **[9 marks] [+ 3 SPaG marks]**

Title of fieldwork enquiry: ....

End of Section B

END OF QUESTIONS

Collins

GCSE
GEOGRAPHY

Resources for Paper 3 Geographical applications

Study the resources in this booklet before completing Paper 3.

The resources for Paper 3 of your actual exam will be issued to you 12 weeks before the date of the exam.

Resource 1: Sources of Energy used to Create Electricity in the UK

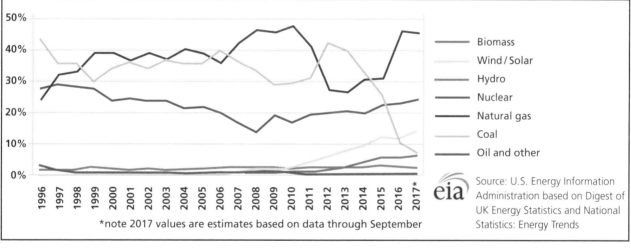

*note 2017 values are estimates based on data through September

Source: U.S. Energy Information Administration based on Digest of UK Energy Statistics and National Statistics: Energy Trends

Resource 2: UK Gas Supplies

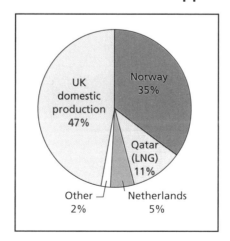

Resource 3: Issue with Power Generation in January 2017

In January 2017, the UK grid faced a perfect storm stemming from nuclear closures in France, nuclear faults in the UK, a broken inter-connector with France and then on 16th January the wind died for a week. These were the exact conditions that were expected to increase the blackout risk, but the lights stayed on.

Concern about the UK grid is borne out of the closure of 17.7 GW of coal-fired power between 2004 and 2016 and its replacement with 14.4 GW of wind and 10.7 GW of solar.

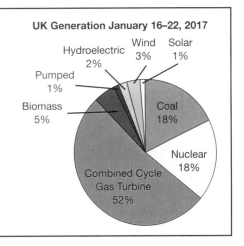

UK Generation January 16–22, 2017

- Hydroelectric 2%
- Wind 3%
- Solar 1%
- Pumped 1%
- Biomass 5%
- Coal 18%
- Nuclear 18%
- Combined Cycle Gas Turbine 52%

Resource 4: Extract from UK Government Statement, April 2021

The UK government will set the world's most ambitious climate change target into law to reduce emissions by 78% by 2035 compared to 1990 levels.

In line with the recommendation from the independent Climate Change Committee, this sixth Carbon Budget limits the volume of greenhouse gases emitted over a 5-year period from 2033 to 2037, taking the UK more than three-quarters of the way to reaching net zero by 2050. The Carbon Budget will ensure Britain remains on track to end its contribution to climate change while remaining consistent with the Paris Agreement temperature goal to limit global warming to well below 2°C and pursue efforts towards 1.5°C.

Resource 5: Photographs of a Wind Farm in the Scottish Highlands

Resource 6: Contribution of Wind Power to Electricity Generation over Time

Year	Capacity (MW)	Generation (GW/h)	% of total electricity use
2011	6,540	12,675	3.81
2012	8,871	20,710	5.52
2013	10,976	24,500	7.39
2014	12,440	28,100	9.30
2015	13,602	40,442	11.0
2016	16,218	37,368	12.0
2017	19,837	49,607	17.0
2018	21,606	56,907	17.1
2019	23,882	63,795	19.7
2020	24,485	75,369	24.1

Resource 7: Taken from a Statement by the John Muir Trust* in May 2018

The news came at the end of April 2018 that Scottish Ministers had refused consent for two major wind farms that the Trust had campaigned against.

Wild Land Area 34: Reay–Cassley, popular walking country and an area of outstandingly significant geology, was threatened by not just one but three industrial-scale wind farms that could have seen a total of 65 turbines radically change this prized landscape.

To our great relief, the Minister rejected two of these concluding that they were simply inappropriate for such an area.

But the third development, which had come into the planning system later, remained a threat until just a few weeks ago when the Minister published his decision following a Public Local Inquiry into this case in 2017, concluding that *"the proposal would not preserve natural beauty … weighing up all of the material considerations, my conclusion is that, on balance, the adverse consequences of the proposal are too significant to be outweighed by its benefits."*

Further south the hilly, moorland country of Wild Land Area 19 has been under threat, from a 13-turbine development that would, according to Scottish Natural Heritage, have led to the loss of seven square kilometres of wild land in the north west corner of the Wild Land Area.

This Wild Land Area is also popular walking country, with five Munros and four Corbetts. It has fantastic high mountains and famous cliffs for winter climbing, and a small overlap with the Cairngorms National Park.

Repeated surveys have demonstrated strong public support for wild land protection.

John Muir Trust is a charity founded in 1983 whose mission is "to conserve and protect wild places with their indigenous animals, plants and soils for the benefit of present and future generations".

Resource 8: Taken from a Statement from the RSPB
(*The Royal Society for the Protection of Birds*)

How do wind farms affect birds?

The available evidence suggests wind farms can harm birds in three possible ways – disturbance, habitat loss and collision.

Some poorly sited wind farms have caused major bird casualties, particularly at Tarifa and Navarra in Spain and the Altamont Pass in California. At these sites, planners failed to consider adequately the likely impact of putting hundreds, or even thousands, of turbines in areas which are important for birds of prey.

If wind farms are located away from major migration routes and important feeding, breeding and roosting areas of those bird species known or suspected to be at risk, it is likely they will have minimal impacts.

Reducing the impact on birds

We are involved in scrutinising hundreds of wind farm applications every year to determine their likely wildlife impacts, and we ultimately object to about 6 per cent of those we engage with, because they threaten bird populations. Where developers are willing to adapt plans to reduce impacts to acceptable levels we withdraw our objections; in other cases we robustly oppose them.

However, there are gaps in knowledge and understanding of the impacts of wind energy, so the environmental impact of operational wind farms needs to be monitored - and policies and practices need to be adaptable, as we learn more about the impacts of wind farms on birds.

Resource 9: Financial Benefits to a Local Community Following Wind Farm Construction

Community Benefit Funds
Funds established by wind farm development companies

One example from the Scottish Highlands

In 2014/15, funds worth approximately £1.5 million were managed.

Some donations in 2016:

- Village Hall: the sum of £7000 to provide free access for local groups in 2016 (following a £3000 donation for 2015)
- The Community Plan Group: £6800 for public access automated defibrillators; these are located at St Paul's Church Hall and four other locations
- Baby and Toddler Group: a total of £81500 for a new play park across from school woods
- Village Summer Activities: the sum of £1409 to help with costs for organising and running the week-long summer school session in the village hall
- £110 towards local events in 2016
- Community Newsletter: a total of £8350 to cover costs
- The Farmers' Association Vintage Rally and Display: the sum of £3429 for 2016
- Shinty Club
- Community Woodlands
- The Music Initiative: £15440 to cover the costs for two terms of music classes
- Indoor Bowls Club: £1000
- Community Hall: a grant of £1600 to cover the costs of new goalposts and for playing fields grass-cutting
- Seniors' Lunch Club: the sum of £2500
- Primary schools and nurseries: £24000 for extracurricular activities during the 2015 school year
- Village Growers: the sum of £7800 towards legal and start-up costs for their community garden project
- Care Group: a total of £9600 in 2016
- Community Access and Transport Association: £11971 towards a new 9-seater bus with wheelchair access
- Reimbursement of ongoing travel costs for those who qualify for free transport

Resource 10: Energy Security

Energy security is the uninterrupted availability of energy sources at an affordable price. In the long term, it attempts to make sure that energy keeps pace with both economic developments and sustainability needs. In the short term, it focuses on the ability of the energy system to react effectively to sudden changes in supply or demand. Lack of energy security has negative economic and social impacts.

Resource 11: UK Wind Power Capability

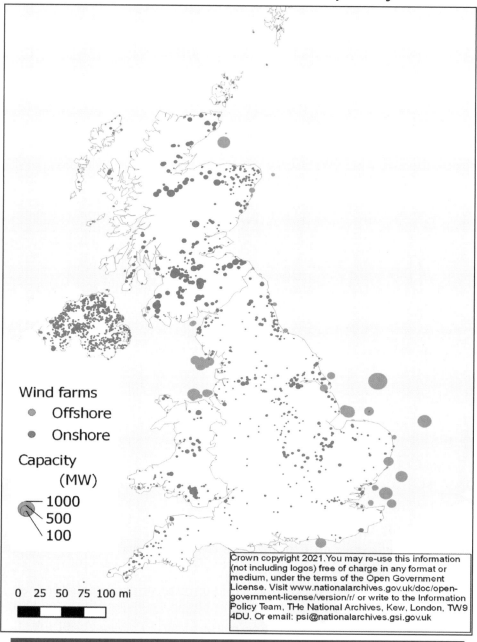

Wind farms
- Offshore
- Onshore

Capacity
 (MW)
— 1000
— 500
— 100

0 25 50 75 100 mi

UK Wind Farm Industry

The UK is one of the best locations for wind power in the world.

Electricity generation from wind power in the UK has increased by 715% from 2009 to 2020.

2020 was the "greenest year on record" for Britain, with record high levels of wind energy generation.

Operational wind farms currently generate

75,610 gigawatt hours

supplying around 24% of electricity to the UK and making the UK the world's sixth largest producer of wind power.

UK onshore and offshore capacity (GW):

10.4

14.1

% of wind farms by country

Answers

TOPIC-BASED QUESTIONS

Pages 148–151: The Challenge of Natural Hazards

1 a) Pacific Ring of Fire [1]
   b) **Examples:** Pacific Plate and North American Plate; Nazca Plate and South American Plate [1]

2 **Example points:** continental plates collide [1]; plates buckle rather than subduct [1]; fold mountains are formed [1], e.g. Alps, Himalayas [1]; no volcanoes [1], but earthquakes occur [1]

3 **Example items:** water [1]; purifying tablets [1]; tinned food [1]; first aid kit [1]; survival blanket [1]; wind-up radio [1]; wind-up torch [1]; batteries [1]. The three named items must be explained as to their usefulness. [3]

4 **Example points:** extreme flooding [1]; difficult to predict [1]; loss of life [1]; damage to coastal settlements and infrastructure [1]; impacts can be at distance from earthquake (e.g. across the Pacific Ocean) [1]; speed of the event [1]

5 **Example points:** homelessness [1]; buildings destroyed [1]; unemployment [1]; disruption to transport [1]; fatalities [1]; shock and trauma [1]; PTSD [1]; out-migration [1]; lack of medical care [1]; lack of education [1]; rebuilding [1]; need for foreign aid [1]

6 **Example points:** magma rising causes land surface to swell [1]; using tiltmeters [1]; measuring chemical composition of gases being emitted [1]; monitoring changes in groundwater temperatures [1]; monitoring earthquake activity [1] with seismometers [1]; satellite and aircraft observations [1]

7 a) Any suitable answer making the key point that the storms are not inland or along the Equator, but over warm ocean water between 5° and 20° north or south.
      Pacific Ocean: occur in the east Pacific but mainly in the west Pacific near the Philippines and north-east Australia; north Atlantic; Indian Ocean (north and south of the Equator); rare in south Atlantic [4]
   b) Areas with higher sea surface temperatures tend to have the most tropical storms [1] but this is not the case along the Equator [1]. They do not form over land [1] and rapidly lose intensity over land because they lose their supply of warm, moist air. [1]

8 **Example points:** affected by the sea [1]; cool, wet winters [1]; warm, dry summers [1]

9 **Example points:** high ground in the west [1] causes heavier relief rain [1] while the east lies in a rain shadow [1], receiving much less rain [1]

10 **Example points:** proxy measures [1]; measure amount of carbon dioxide and other gases [1], trapped in the ice; pollen analysis [1]; show 'ice ages' and interglacials [1]

11 **Example points:**
    Adaptation – adjusting to changes in the environment [1]; building houses on stilts [1]; new agricultural methods [1]; building sea defences [1]
    Mitigation – reducing greenhouse gas emissions [1] and therefore the causes of climate change [1]; carbon capture and storage technology [1]; planting trees [1]; switching to renewable energy [1]; international agreements [1] such as the Paris Agreement [1]

Pages 152–154: The Living World

1 Tourism that uses the beauty of the environment itself as the sole attraction [1] and helps to ensure the protection of the environment [1].

2 **Any answer from:** tropical rainforest; tropical grasslands (savanna); temperate grasslands; temperate forest; boreal forest (taiga); tundra; desert [1]

3 **Example points:** deforestation [1] destroys the habitats of countless organisms, large and small, creating imbalance [1]; draining wetlands [1]; cutting down hedgerows [1]; planting non-native species [1]

4 **Example points:** plants grow [1]; eaten by herbivores [1]; they in turn are eaten by carnivores [1], which die and decompose [1] and the remains are reabsorbed by plants [1]

5 The Poles [1] and the ITCZ (Doldrums, Equator) [1]

6 **Example points:** selective logging [1]; international agreements such as CITES [1]; labelling schemes (FSC) [1]; debt reduction [1]; encouragement of ecotourism [1]; promoting employment [1]; reducing out-migration [1]

7 To provide stability for the very tall trees [1] and to absorb nutrients more easily [1]

8 Evergreen coniferous trees [1]

9 **Example points:** No leaves [1] to reduce transpiration [1] and have needles instead [1]; widespread root system [1]; fleshy leaves and stems [1] to retain moisture [1]

10 **Example points:** land turning to desert [1]; land degradation in dry areas [1]

11 a) A named area must be given. Points may include: increasing temperatures [1] and reduced rainfall [1], as a result of climate change [1]; increased demand for water as a result of increasing population growth [1]; overgrazing by cattle [1]; excessive irrigation of crops [1]
    b) Answer must relate to area studied. Points may include: drip irrigation methods [1] and other named appropriate technologies [1]; increased recycling of water [1]; greater use of greywater [1]; limiting the extent of tourist facilities, including golf courses, water parks, etc. [1]; tree planting [1] because the roots bind the soil together [1] and help to stop the soil being blown away [1]

12 a) **Example points:** greater exploitation of marine resources [1]; fossil fuels extraction [1]; along with other minerals [1]
    b) **Example points:** contravention of existing protocols [1], such as Madrid [1]; overfishing [1]; marine pollution [1]; land pollution (tar sands) [1]; geopolitical tensions [1]

13 **Example points:** the Madrid Protocol is part of the Antarctic Treaty System, signed in 1991 [1] to protect the environment of Antarctica [1]; preserve it as a 'natural reserve, devoted to peace and science' [1]; expressly prohibits mining [1]

14 **Example points:** permafrost is ground that has been frozen for two or more consecutive years [1]; it consists of soil, gravel and sand, usually bound together by ice [1]

15 **Example points:** create attraction of hostile environments to the tourist [1]; includes polar regions, high mountains, deserts [1]; damage to fragile environments and habitats [1]; litter and pollution [1]; too many tourists in a single area [1]; difficulties of international management and agreements [1]

Pages 155–158: Physical Landscapes in the UK

1 **Example points:** water must be present [1]; temperatures need to go above and below 0°C [1]; rocks need to be weakened or cracked [1]

2 **Example points:** rock fragments present [1]; as the ice moves, these fragments scratch the surface of the solid rock [1], wearing it away [1] or leaving scratch marks or striations [1]

3 **Example points:** rounded [1], elongated mounds of moraine [1] shaped by the moving ice [1]. The blunt end faces the approach of the ice [1] and the tail points in the direction of the ice as it flowed over [1]

4 **Example points:** high [1], steep slopes [1]; often impermeable rocks [1]; waterlogged soils [1]; short growing season [1]; windy [1]; short daylight hours [1]

5 **Example points:** impermeable rocks [1]; high rainfall totals [1]; steep slopes [1]; low value land [1]; fast-flowing streams [1]; lakes for reservoirs [1]

6 a) Hard [1]
   b) Rock armour / rip-rap [1]; sea wall [1]

7 A named stretch of coast must be given. Depending on the chosen location, points may include: protection of settlements [1] or infrastructure [1]; sea walls [1]; rock armour [1]; gabions [1]; revetments [1]; groynes [1]; beach feeding [1]

8 **Example points:** longshore drift [1]; supply of sediment [1]; coast changing direction [1], e.g. estuary mouth [1]; onshore winds [1] from two directions to create curved end [1]

9 A named location must be given. Depending on the chosen location, points may include: relatively cheap [1]; cost effective [1]; creates saltmarsh [1]; wildlife habitat [1]; good natural defence for low-lying areas [1]; restores natural sediment movement [1]

10 **Example points:** hydraulic action [1], undermining river banks [1]; abrasion [1]: rock fragments in water wear away at the bed and banks [1]; solution [1]: limestone rocks dissolving in slightly

acidic stream water [1]; attrition [1]: particles hitting each other and reducing in size and becoming rounder [1]

11 **Example points:** sinuous route of water in channel (thalweg) [1]; faster water hits outside of bend [1] and erodes more rapidly [1]; river cliffs form [1]; inside bend slower water deposition [1] forms slip-off slope [1]; have an asymmetrical cross section [1]

12

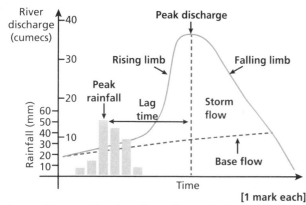

[1 mark each]

13 A named river must be given. Depending on the chosen location, features may include: ox-bow lake, floodplain, levée [4]

14 **Example points:** exceptionally high levels of rainfall [1], in a short period of time [1]; urbanisation brings creation of impermeable surfaces [1], preventing infiltration [1]; upstream changes in land use [1]; deforestation [1]; rapid snow melt [1]; storm surges [1]; exceptionally high tides [1]

15 **Example points:**
Advantages: protection of housing [1], infrastructure [1], railways [1] and electricity supply [1]; can be a tourist attraction [1]; benefits outweigh the costs [1]
Disadvantages: costly to build [1] and maintain [1]; reduces access to the river [1]; traps water should it spill over [1]; unsightly [1]; long-term project [1]; may move flooding problems elsewhere [1]

Pages 159–160: Urban Issues and Challenges
1 Where urban areas grow outwards [1] into the surrounding rural regions [1]
2 **Example points:** rural-to-urban migration [1]; high natural increase [1]; high birth rates and falling death rates [1]; better opportunities [1]
3 **Example points:** high infant mortality rates [1]; lack of medical services [1]
4 **Example points:**
Push: un/underemployment [1]; natural disasters [1]; low wages [1]; difficult farming conditions [1]; lack of alternative job opportunities [1]; isolation [1]; lack of social amenities [1]; lack of schools, hospitals, etc. [1]
Pull: perception of [1] job opportunities [1]; better wages [1]; access to schools, hospitals, etc. [1]; access to transport [1]; better housing conditions [1]; better access to water and electricity [1]; more social amenities [1]
5 a) Specific challenges will depend on the city chosen. Points may include: in-migration from rural areas [1]; informal housing [1]; insufficient healthcare [1]; insufficient education [1]; lack of clean water supply [1]; shortage of electricity [1]; crime [1]; poverty [1]; un/underemployment [1]; pollution [1]; waste disposal [1]
b) Specific solutions will depend on the city chosen and must match with the challenges given in part a). Points may include: upgrading of informal housing [1]; self-build schemes [1]; water and electricity provision [1]; legal rights to land ownership [1]; improved transportation systems [1]; law and order improved [1]; tourism encouraged [1]
6 Specific opportunities will depend on the city chosen. Points may include: redevelopment of old brownfield sites [1]; creation of new jobs in high tech industries [1]; transport improvements [1] (giving a specific example [1]); promotion of tourism [1]; regeneration projects [1] (giving a specific example [1])
7 Specific challenges will depend on the city chosen. Points may include: housing inequalities [1]; variations in educational achievement [1]; in-migration [1]; loss of secondary jobs in industrial areas [1], leading to unemployment [1]; pay inequality [1]; dereliction in old industrial areas [1]; pollution [1]

8 **Example points:** affordable house prices/rents [1]; shared housing [1]; natural lighting where possible [1]; passive heating wherever possible [1]; recycling [1]; triple-glazed windows [1]; renewable energy use [1]; water meters [1]
9 Examples must be given with both named strategies. Points may include: car sharing [1]; pedestrian-only areas [1]; street closures [1]; integrated public transport systems [1]; park-and-ride schemes [1]; bus lanes [1]; encouragement of cycling [1]; dedicated pedestrian walkways [1]
10 **Example points:** vertical farming / allotments / rooftop gardens [1]; locally grown produce [1] is sold locally [1]; removes need for long distance transportation [1]
11 **Example points:** aesthetically pleasing (looks nice) [1]; improves well-being / mental health [1]; provides shade on streets [1]; helps reduce air and noise pollution [1]; trees take in carbon dioxide [1]; provides oxygen from photosynthesis [1]; intercepts / soaks up rainfall/water [1]; helps to prevent flooding [1]

Pages 161–163: The Changing Economic World
1 **Example points:** data given is for the whole country [1]; some regions within a country might be more developed than others [1] and the figure given is only an average [1]
2 **Example measures:** life expectancy [1]; GNI [1]; GDP [1]; literacy rate [1]; birth rate [1]; death rate [1]; people per doctor [1]; calorie intake [1]; car ownership [1]; access to safe water [1]; infant mortality rate [1]
3 Population changes over time [1], which can be related to stages of development [1]
4 **Example points:** climatic extremes [1]; unproductive land for farming [1]; unreliable water supplies [1]; natural hazards [1]; limited natural resources [1]
5 **Example points:** disparity in wealth distribution [1]; disparities in health and medical services [1]; international migration [1]
6 **Example points:** appropriate simple technology [1] as the most up-to-date might not be suitable to the conditions [1]; specific example [1]
7 **Example points:** where raw material producers are paid a fair price for their goods [1] such as Fair Trade bananas/coffee/cotton [1]; allows farmers to plan ahead [1], knowing their income is secure [1]
8 A named LIC or NEE must be given. Points may include: can account for large percentage of GDP [1]; local cultures are preserved [1]; establishment of national parks protect wildlife and landscapes [1]; improvements to infrastructure for visitors [1] also benefits local people [1]; improved quality of life [1] as a result of increased levels of income [1]
9 A named example must be given. Points may include:
Advantages: improved levels of wages [1]; increased levels of female employment [1]; new skills [1]; increased levels of tax paid [1]; allows greater spend by government on infrastructure [1]; attracts other TNCs [1]
Disadvantages: national cultures can be undermined [1], by westernisation [1]; TNC may exert 'control' over government decision making [1]; environmental impacts / pollution [1]; excessive use of water [1]; TNC could move away at short notice [1]
10 A named country must be given. Points may include:
Short-term aid: to provide assistance following a natural disaster [1]; providing temporary housing [1]; medical support [1]; repair to infrastructure/roads/airport [1]
Long-term aid: major project such as building a dam/hospitals/airport [1]; developing the education/medical infrastructure [1]; developing natural resources [1]
11 **Example points:** traditionally, the north has heavy industry [1] and the south has services and high tech industries [1]; wages higher in the south [1]; unemployment higher in the north [1]; house prices higher in the south [1]; students have better exam results in the south [1]; life expectancy is longer in the south [1]
12 **Example points:** when primary (raw materials) and secondary (manufacturing) industries have declined [1], as a result of mechanisation and automation [1], exhaustion of local raw materials [1] and/or competition from overseas [1]
13 **Example points:** use of modern technology to reduce emissions [1]; named example [1]; desulphurisation [1]; tighter controls on water pollution [1]; restoration of damaged landscapes [1]; fines for pollution [1]; greater use of old brownfield sites for future development [1]
14 **Example points:** loss/closure of shops [1] and other local amenities such as primary school [1], doctor [1], church [1] and bus

services [1]; out-migration of younger people [1]; in-migration of second homers and weekenders [1]

15 Named examples must be given. Points may include:
Social benefits: people moving away from crowded urban areas [1], such as the south-east [1]; less stressful journeys to work [1]; better national links [1]; easier access to airports [1]
Economic benefits: job creation [1]; increases people's access to job opportunities [1]; increased speed of business deliveries [1]; encourages business investment [1]; can lead to a positive multiplier effect [1]

Pages 164–167: The Challenge of Resource Management

1 The amount of carbon dioxide released into the atmosphere [1] as a result of the activities of a particular individual, organisation, or community. [1]

2 **Example points:** only 25% of our food is grown in the UK [1], so we import from far afield [1]; demand is for crops 'out of season' in UK [1], e.g. asparagus from Peru [1]

3 **Example points:** increasing affluence [1]; increasing population [1]; two-bathroom houses [1]; more appliances, e.g. washing machines [1]

4 **Example points:** highest demand in the driest part of the country [1], i.e. London and the south-east [1]; least demand in the wettest part of the country [1], i.e. the north and west [1]; water sent by pipeline and river [1], e.g. Thirlmere to Manchester [1]; London faces significant problems in the future as it is so far from supplies [1]

5 Fossil fuels are running out [1]; less than 40% (2021) from renewables [1], although increasing [1]; oil, gas etc. imported [1]; foreign supplies not guaranteed [1]; countries moving away from Russian supplies following the invasion of Ukraine in 2022 [1]

6 **Example points:** There are many different groups of countries: some, like those in the Middle East and North Africa, are arid and have shortages of water [1]. Others, like South Africa and India [1], have poor infrastructure and are not able to maintain supplies to their population [1]. Other countries, like the UK and China [1], have large population densities [1] and water in some areas may be in short supply due to high demand [1].

7 Energy generated using resources that can be used over and over again [1]; examples include solar, wind, wave, tidal, hydro. [2]

8 **Example points:** noise [1]; increased traffic pollution [1]; risk of 'earthquakes' [1]; pollution from spillages [1]; environmental damage [1]; aesthetically damaging [1]; reduced house prices [1]

9 **Example points:**
Physical reasons: water availability [1]; high temperature [1]; poor soils [1]
Human reasons: population size [1]; skill levels [1]; lack of investment [1]; emphasis on producing export crops [1]

10 **Example points:** extreme climatic conditions [1], such as drought [1] or other natural disaster example [1]; low yields [1]; environmental degradation [1]; soil erosion [1]; desertification [1]; social unrest / civil war [1]

11 The application of water to increase yields in areas where water is in short supply [1]

12 **Example points:** development of new crops that produce higher yields [1], such as IR8 (a variety of high-yielding rice) in India [1]; development of crops that need fewer chemical inputs [1]; farmers able to afford to grow them [1]

13 Named example must be given. Points may include:
producers grow for their own use [1]; vegetables, fruit, bees, etc. [1]; surplus can be sold locally [1]; small-scale drip irrigation methods [1]; organic waste is recycled to enrich the soil [1]

14 Named examples must be given. Points may include:
water transfer systems [1]; desalinisation plants [1]; water conservation [1], e.g. two-flush toilets [1]; use of greywater for toilets [1]; drought-resistant plants in gardens [1]; water butts used rather than hoses [1]

15 **Example points:** increased population means greater demand [1]; increased economic development leads to increased consumption [1], both domestic and industrial [1]; affluence means higher living standards [1], including more electrical items and gadgets [1]

16 **Example points:** intelligent building design [1]; insulation / double glazing to prevent heat loss [1]; energy-efficient devices [1]; auto switch-off features [1]; LED light bulbs [1]; cold-wash washing machines [1]; smart meters to control temperatures [1]; encouraging use of public transport rather than cars [1]

Page 168: Geographical Applications

1. Answers will vary. [8]

PRACTICE EXAM PAPERS
How exam papers are marked
Point marking

1, 2 and 3-mark questions are point marked. This means that if you give a correct answer, you attain a mark. For 2 and 3-mark questions, you can often earn a mark by developing your answer. For example, an answer of traffic congestion would earn 1 mark, but by linking it to longer journey times or delays to deliveries would earn an extra development mark.

Level marking

4, 6 and 9-mark questions are level marked. Each level has a description. The marker will read your answer and determine its level. Once they have assigned a level, they will then decide on a mark. The table below shows you some outline descriptors for each of the different levels.

Type of Question	Level 1 – Basic	Level 2 – Clear	Level 3 – Detailed
4 marks	Limited understanding. Limited/Unclear use of the figure. Limited application of knowledge and understanding.	Clear understanding. Effective use of the figure. Applies knowledge and understanding. Answers all aspects of the question.	
6 marks	Limited understanding. Limited application of knowledge and understanding. Limited use of the figure and/or a named example or case study.	Clear understanding. Some application of knowledge and understanding. Some use of the figure. Uses a named example or case study.	Detailed understanding. Thorough application of knowledge and understanding. Clearly uses the figure. Detailed use of a named example or case study. Answers all aspects of the question.
9 marks	Limited knowledge. Limited understanding. Limited application of knowledge and understanding. Limited use of the figure and/or a named example or case study.	Reasonable knowledge. Clear understanding. Reasonable application of knowledge and understanding. Some use of the figure. Some use of a named example or case study.	Detailed knowledge. Thorough understanding. Thorough application of knowledge and understanding. Accurate use of the figure. Detailed use of a named example or case study. Answers all aspects of the question. Draws a considered conclusion.

Assessment of spelling, punctuation, grammar and use of specialist terminology (SPaG)

Accuracy of spelling, punctuation, grammar and the use of specialist terminology will be assessed when indicated beside a 9-mark question. The cover sheet of each exam paper identifies which questions will be used. In each of these questions, 3 marks are allocated for SPaG as follows:

High performance – 3 marks
- spelling and punctuation is consistently accurate
- well-written with excellent use of grammar, so meaning is always clear
- a wide range of specialist terms are used appropriately

Intermediate performance – 2 marks
- spelling and punctuation is considerably accurate

- good use of grammar, so meaning is generally clear
- a good range of specialist terms are used appropriately

Threshold performance – 1 mark
- spelling and punctuation is reasonably accurate
- reasonable use of grammar (overall, any errors do not significantly hinder understanding of the answer)
- a limited range of specialist terms are used appropriately

No marks awarded – 0 marks
- no answer has been given
- the learner's response does not relate to the question
- spelling, punctuation and grammar does not reach the threshold level (errors mean that the answer cannot be properly understood)

For the purpose of this book, answers have been provided for 1, 2 and 3-mark questions so that you can mark your own answers. For level-marked 4, 6 and 9-mark questions, key points that should be addressed have been provided along with a model answer so that you can see what a good answer looks like. It is important to remember that you will only be awarded full marks in the exam if these points are communicated in a clear, accurate and well-developed way.

Pages 169–190
Paper 1: Living with the physical environment
Section A: The challenge of natural hazards

01.1 C **[1]**

01.2 Destructive plate margin / Convergent plate margin **[1]**

01.3 Your answer must refer to both places to show the contrast.
Example:
The UK is well within a tectonic plate / the Eurasian Plate, so shockwaves have further to travel / will be absorbed before affecting the UK **[1]** but three tectonic plates (may be named) meet at Japan **[1]**.

01.4 This question is level marked. Your answer must:
- refer to Figure 2
- cover both primary and secondary effects
- include a range of ideas.
You could include primary and secondary effects from either a volcano or earthquake event that you have studied.
Example points:
In Figure 2, ground shaking from the earthquake looks like it has damaged some buildings, particularly those to the right of the photograph. This is a primary effect. These homes may not be safe to live in, causing people to temporarily live somewhere else. This may cause stress to the residents. In the Eyjafjallajökull volcano eruption in Iceland, ash affected local farms and caused the loss of some livestock. Ash also caused the cancellation of thousands of European flights, leading to travellers being stranded across Europe and a loss of business to some companies. In Figure 2, it looks like the ground shaking has caused soil liquefaction. This is a secondary effect and has caused damage to roads, which may affect accessibility in the area. In Iceland, the publicity created by the volcano eruption has led to a long-term increase in tourism, which has created jobs and provided more income to some businesses.

01.5 B **[1]** and D **[1]**

01.6 Winds blow from areas of high pressure to areas of low pressure **[1]**

01.7 This question is level marked. Your answer must:
- show a clear understanding of the causes of tropical storms
- include several developed explanations.
Example points:
Tropical storms form over warm oceans between 5° and 20° north and south of the Equator. Sea surface temperature must be 27°C or more. The warm oceans cause air to rise; this is low pressure. As the air rises, it cools and condenses forming large, towering clouds leading to convectional rain. The Coriolis Force causes a tropical storm to rotate, anticlockwise in the northern hemisphere and clockwise in the southern hemisphere.

01.8 You could agree or disagree with the statement depending on the evidence you use.
Example points:
There is a link between maximum wind speed and the number of deaths. The two tropical storms with the highest maximum wind speeds caused the most deaths **[1]**. Hurricane Katrina with a wind speed of 175 mph caused 1200 deaths; this is four times the number of deaths caused by Superstorm Sandy, which had the lowest maximum wind speed at 110 mph **[1]**.

There is no clear link. Some of the storms with the highest wind speeds caused low number of deaths **[1]**. For example, Hurricane Michael had faster wind speeds than Superstorm Sandy, but Superstorm Sandy caused approximately six times more deaths **[1]**.

01.9 They may affect areas further away from the Equator / They may affect larger parts of the world **[1]**. They will become more intense / frequent **[1]**.

01.10 This question is level marked. There are also 3 marks for spelling, punctuation and grammar. Your answer must:
- refer to Figure 5
- cover both mitigation and adaptation strategies
- include a range of well explained ideas
- include ideas related to mitigation and adaptation from your own studies
- use named examples
- include a conclusion where you summarise your overall answer to the question.
Example points:
Figure 5 shows alternative energy production strategies, such as solar panels on house roofs and wind turbines. These are renewable energy sources and, unlike fossil fuels, don't release carbon dioxide into the atmosphere and therefore help mitigate against climate change. Many countries like the UK are trying to develop cleaner sources of energy. Another mitigation strategy shown is planting trees. Trees are carbon sinks. They take in carbon dioxide through photosynthesis, and this reduces the amount in the atmosphere, therefore helping combat climate change. Mitigation strategies are an important way of managing climate change as they stop the causes of climate change by reducing the amount of greenhouse gases in the atmosphere. In countries such as Bangladesh, they are adapting to the effects of climate change by building houses on stilts so that they are not affected by rising sea levels. This helps to reduce damage to people's property and possessions. Figure 5 shows farming in what looks like very dry conditions. In countries such as Burkina Faso, they are growing crops that are drought resistant. This enables farmers to be able to continue to provide food and earn a living despite rising temperatures and longer periods of drought caused by climate change. In places such as Cape Town in South Africa, they are investing in water supply strategies like desalination to ensure water security even when there are extreme droughts. In conclusion, both adaptation and mitigation strategies are needed to manage climate change. Adaptation enables us to continue to live within a changing environment, whilst mitigation strategies are needed to limit the amount that the climate changes.

Section B: The living world

02.1 B **[1]**

02.2 B **[1]**

02.3 A community of plants and animals / A community of living and non-living elements **[1]**

02.4 Decomposers help to return nutrients/energy to the soil / Decomposers break down dead plants and animals or excreted material **[1]**

02.5 **Any one from:** Insect; rabbit; mouse; red deer; bird; hare **[1]**

02.6 Fewer red deer means less demand on plants **[1]**, so more of those are available to other primary consumers, e.g. rabbits, insects **[1]**. The numbers of other primary consumers might increase as a result of more food being available **[1]**. Less food available for eagles **[1]** because they are the only predator of red deer **[1]**. Eagle numbers might decrease **[1]**. Eagles would hunt more rabbits, mice or hares **[1]**, whose numbers could decline **[1]**.

02.7 Rainfall occurs all year round **[1]**. There are seasonal patterns of rainfall (named months may be given) **[1]**. The highest rainfall, 295 mm, occurs in March **[1]**. The lowest rainfall, 55 mm, occurs in August **[1]**. The range in rainfall is 240 mm **[1]**.

02.8 The sun is much higher in the sky for most of the year **[1]** / The sun's rays have a relatively short distance to travel **[1]** / The sun's rays are more concentrated **[1]**

02.9 This question is level marked. Your answer must:
- refer to Figure 9
- cover both economic and environmental impacts
- include a range of ideas.

You could include ideas from a case study you have studied.

Example points:

Figure 9 shows an area of logging and mining in the Amazon Rainforest. These economic activities provide jobs to local people. This can lead to a positive multiplier effect because workers spend their earnings in local businesses such as shops, which boosts the local economy. Furthermore, taxes paid by workers to the government enable them to invest more in services such as health and education. A lot of the resources extracted are exported to other countries, which helps to boost trade. For example, a lot of copper that is mined in the Amazon Rainforest is exported to China. In terms of environmental impacts, deforestation can lead to climate change. Trees are a carbon sink, so cutting them down means that photosynthesis cannot take place, therefore less carbon dioxide is taken out of the atmosphere. Soil erosion can also be an issue. Once the trees are removed, the soil is left exposed and there are no roots to hold the soil together. Heavy rain can wash the top soil away, removing nutrients. This can silt up rivers, which can lead to flooding.

02.10 This question is level marked. Your answer must:
- use a case study
- cover both plants and animals
- include a range of well explained ideas
- include a conclusion where you summarise your overall answer to the question.

Example points:

Hot desert environments (such as the Mojave Desert in USA or the Thar Desert in India)

Plants have no leaves or small seasonal leaves that only grow after it rains; this helps to reduce water loss during photosynthesis. Plants such as acacia trees have long root systems. These spread out wide or go deep into the ground to absorb water. Plants such as cacti have spikes instead of leaves; this reduces water loss and protects the plant from being eaten by animals. Many plants have a waxy coating on stems and leaves; again this helps to reduce water loss.

Animals like the fennec fox have large ears to give off heat. Many animals produce little urine to save water and are active only at night to avoid the heat. Camels have two sets of eye lashes to keep sand out of their eyes. Desert foxes have light coloured fur to reflect heat and thick fur on the soles of their feet, protecting them from the hot ground.

Cold environment (such as Alaska in USA)

Plants grow close together and low to the ground. Plants have shallow root systems to access meltwater produced by the top layer of the permafrost melting. Some plants have a waxy, hairy coating which helps to shield them from the cold and the wind and protects plant seeds. They have small leaves, which helps the plants to retain moisture. Most plants don't die off in the winter; they have long life cycles to help with the short growing season. This means photosynthesis can begin immediately once the sunlight is strong enough as the plants don't need to regrow leaves. Plants like the Arctic poppy flower have cup-shaped flowers that face the sun to capture as much sunlight as possible.

Animals tend to have thicker and warmer fur and feathers. Many of them have larger bodies and shorter arms, legs and tails, helping them to retain heat. Animals like polar bears and the Arctic fox have thick fur and feet lined with fur to help keep them warm. The Arctic hare has small ears to reduce heat loss and white fur to avoid the gaze of predators. Some animals, like Arctic squirrels, hibernate for the winter.

In conclusion, plants and animals have adapted to a great extent to survive the harsh physical conditions of a hot desert environment / cold environment.

Section C: Physical landscapes in the UK

03.1 C **[1]**

03.2 D **[1]**

03.3 An answer between 150 and 200 m **[1]**

03.4 The area has a camping and caravan site / holiday park **[1]**, which shows that people visit the area and stay overnight **[1]** / There is a picnic site **[1]** and this shows that people visit the area **[1]**. Answers may also refer to the stained glass centre, public houses and long distance footpath as evidence of the area attracting visitors.

03.5 This question is level marked. Your answer must include a range of well explained ideas.

Example points:

Sand dunes are created by deposition. Constructive waves deposit sand on to a beach because the swash of the waves is greater than the backwash. Onshore winds carry the sand towards the back of the beach. Sand will build up next to obstacles such as driftwood. Vegetation such as sea rocket colonise the sand dunes, which stabilise them and allow the landform to develop.

03.6 This question is level marked. Your answer must:
- refer to Figure 11
- cover both costs and benefits
- include a range of ideas.

You could include ideas from a coastal management example that you have studied.

Example points:

Figure 11 shows a sea wall, rock armour and groynes. These are man-made coastal management strategies. One benefit of these strategies is that they reduce rates of coastal erosion by absorbing wave energy, and they can protect land from flooding. These methods are generally effective and can last several decades. A cost of these strategies is that they are artificial and often don't look very nice on the landscape. They often cost a lot of money. For example, sea wall costs £5000 a metre. Groynes can restrict the supply of sediment further down the coast. For example, villages south of Mappleton along the Holderness Coast in Yorkshire have experienced greater rates of erosion. This has led to homes and land being lost with resulting economic losses.

04.1 A **[1]**

04.2 Land is very steep **[1]** / The river is in a valley **[1]**

04.3 B **[1]**

04.4 Vertical erosion / the river cutting down **[1]**; Hydraulic action / the power of water wearing away the bed and the banks of the river **[1]**; Abrasion / material within the river hitting or being scraped against the bed and banks **[1]**; Solution / acids in the water reacting with certain types of rock, such as limestone **[1]**.

04.5 This question is level marked. Your answer must include a range of well explained ideas.

Example points:

Levées are formed by deposition and are features that are found in the middle and lower courses of a river. When a river floods, it bursts its banks. Friction with the land causes the velocity of the water to slow down, so it deposits its load. Heavy material such as gravel is deposited first and is found closest to the river. The size of sediment becomes smaller with increased distance from the river. With each flood, the riverbanks are built a little higher. These raised banks are levées.

04.6 This question is level marked. Your answer must:
- refer to Figure 13
- include ideas from an example of a flood management scheme you have studied
- cover social, economic and environmental issues
- include a range of ideas.

Example points:

Figure 13 shows tree planting and river embankments. An economic issue is that both these management methods cost money. River embankments more so than tree planting as they also require maintenance. That said, compared to management strategies such as dams and reservoirs they are relatively cheap. Both management methods can reduce flood risk, meaning there is less economic damage to property and infrastructure; this can help reduce insurance costs. Socially, embankments such as those along the River Thames in London, provide walkways and cycle paths for people to enjoy. Embankments can, however, reduce access to the river for boats and fishing. Environmentally, soft engineering methods like tree planting can create habitat for wildlife. Furthermore, trees are a carbon sink, absorbing carbon dioxide and helping to mitigate climate change. Hard engineering methods like embankments can disfigure the landscape and their installation can destroy habitats. This was the case with the Jubilee flood channel in south-east England.

05.1 A **[1]**

05.2 893 − 766 = 127 m **[1]**

05.3 Much of the land is either open space or agriculture **[1]**. There is some woodland in the south-east of the area **[1]**. A footpath crosses the area **[1]**. A road / track passes through the north-west of the area. Some of the area is occupied by a lake (Llyn Mwyngil) **[1]**.

05.4 A **[1]**

05.5 This question is level marked. Your answer must include a range of well explained ideas.

Example points:
A drumlin is a landform created by ice transport and deposition. They are long, smooth, rounded mounds of moraine often found in clusters. Drumlins are made up of glacial material eroded by the glacier further up the valley. Firstly, material is deposited under a glacier. Melting ice at the base of the glacier causes material to be deposited as ground moraine, as there is too much to be carried. This ground moraine is then sculpted to form drumlin shapes by further ice movements. They usually have a blunt end that faces up the valley and a more tapered end facing down-valley.

05.6 This question is level marked. Your answer must:
- refer to Figure 15
- include ideas from an example of a glaciated upland area used for tourism in the UK you have studied
- include a range of ideas.

Example points:
Figure 15 shows an information board. These are often used in areas like the Lake District (in England) and Snowdonia (in Wales) to encourage people to stay on footpaths. This can help to reduce erosion of the landscape; however, these signposts are often ignored by people. In the Lake District, a 'Fix the Fells' scheme is run. This repairs and maintains footpaths. Where possible, natural materials are used and native species planted so that the area continues to look visually pleasing. The scheme relies on volunteers and donations, so work is often slow, and many footpaths are still in need of attention. Figure 15 also shows a bus, possibly part of a park-and-ride scheme. This encourages visitors to use public transport rather than their cars, reducing noise, air pollution and congestion on local roads.

Pages 191–212
Paper 2 Challenges in the human environment
Section A: Urban issues and challenges

01.1 C **[1]**

01.2 0 to 0.9% **[1]**

01.3 There is better healthcare **[1]** so the death rate is lower than the birth rate, leading to natural increase **[1]**. The birth rate is higher in cities **[1]** due to the large number of young adults that live there **[1]**. The population is youthful **[1]** so they are more likely to have children, which increases the population **[1]**.

01.4 Disease can spread quickly **[1]** as people live in high densities **[1]**. Building houses is difficult **[1]** owing to the steep relief of the land **[1]**. Services such as education and health can be overstretched **[1]** owing to the high numbers of people that live in the area **[1]**. Buildings can collapse **[1]** as heavy rain on the steep slopes can cause mudslides **[1]**.

01.5 This question is level marked. Your answer must:
- refer to an example that you have studied
- include a range of ideas.

Example points:
In Rocinha, a shanty town area of Rio de Janeiro in Brazil, various projects have been put in place to improve quality of life. The Favela Bairro Project led to the upgrading of favelas. Homes were connected to water, electricity and sewerage systems. This has improved quality of life by giving residents easy access to clean water, which has helped improve health. It has also reduced the time spent walking to collect water, meaning that people have more time for work or education. Building materials have been improved, with bricks and concrete used rather than wood and scrap materials. This makes buildings permanent and less likely to be damaged by extreme weather. Furthermore, residents can apply to own their properties legally, which encourages them to improve their houses. This is called self-help. Improved transport systems, such as the cable car system in Complexo do Alemao, has made it easier for people to travel to the city to get work.

This improves quality of life by enabling people to earn money, which they can then spend on things such as food, healthcare and household improvements.

01.6 Providing green spaces in urban areas **[1]** / Planting trees, and building parks and gardens in cities **[1]**.

01.7 This question is level marked. Your answer must:
- refer to Figure 3
- include several developed explanations.

Example points:
In Figure 3 there are lots of plants on the side of the building and around balconies. These plants will absorb carbon dioxide, helping to reduce carbon emissions and improving air quality. Furthermore, the plants look nice and improve the view for residents and workers in the building. Trees have also been planted near to the building. This will help provide shade so reducing the need for energy for air-conditioning.

01.8 **Example:** London; south-east England **[1]**; capital city **[1]**; location of central government **[1]**; centre of finance **[1]**; lots of international transport hubs **[1]**; centre of tourism **[1]**

01.9 In London, migrants have helped to fill jobs where there are shortages **[1]** and the mainly young migrants have helped to balance the ageing population **[1]**. Through paying taxes, migrants contribute to the economy **[1]** and they can create a multicultural population, enriching the city's cultural life **[1]**. Migrants have added pressure to housing **[1]**, leading to increases in price **[1]**. Challenges have been created in health and education services where people's first language is not English **[1]**.

01.10 This question is level marked. There are also 3 marks for spelling, punctuation and grammar. Your answer must:
- refer to Figure 4
- include a range of well explained ideas
- use a named example
- recognise that urban regeneration can bring both positive and negative impacts
- include a conclusion where you summarise your overall answer to the question.

Example points:
Figure 4 shows Ancoats, an area of Manchester. Before regeneration there is a building that is boarded up and empty. The paintwork on the building is dirty and the building has graffiti on the wall. People that worked in this building have probably lost their jobs and the area does not look nice. After regeneration, the area looks more welcoming. There are trees and baskets of flowers that make the area look more attractive. Plants will also absorb pollutants from traffic, which help to improve air quality. The building to the right looks like flats, which will provide housing for people and perhaps create a new community. However, if the flats are high priced, like those in London Docklands, they will be too expensive for local people to afford and flats are not suitable for families. This may push people out of the area. The building to the left looks it has some businesses in it. This could provide jobs and attract visitors to the area, creating a positive multiplier effect. In London Docklands, lots of new office-based jobs have been created at Canary Wharf; however, many of these jobs are finance-related and not suitable for the skills of local people. Developments such as Queen Elizabeth Park and the O2 Arena have created new cultural opportunities for local people and visitors. In conclusion, regeneration can solve urban problems. However, to be completely successful, existing residents' needs have to be met.

Section B: The changing economic world

02.1 9070 **[2]**
[1 mark for values put in order; 1 mark for correct answer. 2 marks if answer is correct with no working out]

02.2 Adult literacy **[1]**; birth rate **[1]**; death rate **[1]**; life expectancy **[1]**; number of people per doctor **[1]**; infant mortality **[1]**; calorie intake **[1]**; access to safe water **[1]**

02.3 Economic measures for LICs are often not accurate **[1]** owing to limitations in data collection **[1]** and the fact that many people either work informally/illegally or as subsistence farmers **[1]**. Mean figures are often not representative of the whole country **[1]**. Economic measures don't take account of social factors **[1]**, which are an important part of people's quality of life **[1]**. Economic measures can be misleading **[1]**. For example, in Saudi Arabia, owing to the oil industry a small number of people are rich **[1]** and this increases GNI per capita **[1]**, but many people live poorly **[1]**.

02.4 This question is level marked. Your answer must include several developed explanations.

Example points:

Fair trade means that farmers get a fair, guaranteed price for the products they produce. A fair price is one that covers the costs of production and enables people to have a reasonable standard of living. This helps to reduce the development gap. Money is often invested into community projects, such as building wells. This helps to reduce the development gap by giving people access to clean water, which improves health and therefore life expectancy. It also enables people to spend more time working, boosting their earnings, rather than walking to collect water. This means people can pay for health and education, again reducing the development gap.

02.5 **Example:** Vietnam. It has a range of natural resources **[1]**, such as coal, timber and offshore oil and gas **[1]**. The climate in the north of the country is monsoon **[1]** with a wet season between May and September **[1]**. In the south, the climate is tropical **[1]**, hot and wet all year round **[1]**.

02.6 This question is level marked. Your answer must:
- refer to a case study that you have studied
- cover both advantages and disadvantages
- include a range of ideas.

Example points:

Cheap labour has attracted many transnational corporations (TNCs) to invest in Vietnam, such as Nike. One advantage is that this has created jobs for local people. The wages paid are often above the minimum wage. Taxes on wages increase the national government's revenue, which can then be invested into building schools, health centres and transport infrastructure. This improves the development of the country, increasing life expectancy and adult literacy. A disadvantage is that the profits made by Nike are returned to the headquarters of the business in the USA. This means that money leaks out of Vietnam's economy. Furthermore, there have been reports of poor conditions for the workers (often young women). A TNC could leave a host country abruptly, causing problems for the local economy.

02.7 B **[1]** and C **[1]**

02.8 This question is level marked. Your answer must:
- include a range of well explained ideas
- use named examples
- recognise that some strategies have been more successful than others
- include a conclusion where you summarise your overall answer to the question.

Example points:

The government is using various strategies to reduce the UK's north-south divide. One strategy is the building of high-speed railways, connecting cities such as Birmingham and Manchester to London. This will reduce journey times between cities and encourage businesses to invest in the Midlands and northern England, creating jobs and providing a boost to their economies. At this time, the success is unknown as the rail link is not fully built. There is concern that it will just create more jobs in London as it becomes quicker to commute there. Another issue is the costs, which are rising and have led to the rail link no longer going to Leeds. Another strategy is the creation of the 'Great Northern Powerhouse' to encourage industrial development in northern cities such as Manchester, Leeds and Sheffield. The idea is that this will create jobs for local people, leading to a positive multiplier effect. Part of this strategy is devolution measures, which gives additional power and money to councils in the north. An example is the election of the Mayor for Greater Manchester, who has been given £1 billion of funds to spend on improving the city and attracting new businesses. This has had some success, for example with the extension of the Metrolink tram system. Devolution means that fewer decisions about Manchester are made in London and more decisions are made by local people. In conclusion, there are a variety of strategies to reduce the north-south divide and, whilst there has been some success with these strategies, the gap still remains.

Section C: The challenge of resource management

03.1 2180 kcal **[1]**

03.2 Low caloric intake can lead to malnutrition **[1]**. This can make people more susceptible to illness and disease **[1]**, which can reduce life expectancy **[1]**. Very high caloric intake can cause obesity **[1]**, making people more vulnerable to disease **[1]**; this lowers well-being **[1]**. People need a certain number of calories to provide them with enough energy **[1]**; this enables people to work and earn money **[1]**, improving well-being **[1]**.

03.3 Higher food prices **[1]** owing to higher transport costs **[1]**. Greater food miles **[1]**, which means more carbon dioxide emissions, causing climate change **[1]**. It could take trade away from domestic farmers **[1]**, which decreases their incomes **[1]**.

03.4 This question is level marked. Your answer must:
- refer to Figure 8
- include a range of ideas.

Example points:

The UK population is increasing and this is leading to higher demand for water by both households and agriculture. As standard of living rises and demand for products such as dishwashers rises, this further increases the demand for water. Figure 8 shows that the highest levels of water stress are in the south-east of England. The south-east of England is one of the driest areas of the country, which means there is a limited supply of water. The high demand from the large population and limited supply leads to a water deficit. In extremely dry years (such as 2006), this can lead to the introduction of water restrictions, such as hosepipe bans. Climate change could lead to this type of situation increasing in the future. In Figure 8, areas like the north-west of England have low levels of water stress. This is because the population of this region is lower and the climate is wetter, leading to a water surplus. In the UK, water is transferred from surplus areas to deficit areas to help to reduce water stress.

04.1 **Any one from:** South Africa; Gabon; Ghana; Mali; Morocco; Algeria; Tunisia; Egypt **[1]**

04.2 A **[1]**

04.3 The majority are in North Africa **[1]**; two are located in east Africa **[1]**; one is located in southern Africa **[1]**; most are found next to the coast **[1]**

04.4 The climate of a country will affect how much food a country can grow **[1]**. For example, arid climates such as those in Egypt will limit the types of crop that they are able to grow **[1]**. Some countries can import food **[1]**, which means they can still access food that they cannot grow themselves **[1]**. Many lower income countries lack the money to invest in agricultural infrastructure, such as irrigation systems **[1]**; this can reduce yields and therefore food supplies **[1]**.

04.5 Not being able to access a reliable and affordable supply of nutritious food **[1]**; a lack of food **[1]**

04.6 This question is level marked. Your answer must:
- refer to Figure 10a and 10b
- include a range of well explained ideas.

Example points:

Figure 10a shows the use of hydroponics to grow crops such as lettuces. Hydroponics is where plants are grown in a nutrient-rich water. This means they receive more nutrients, which leads to faster growth. This method uses less space as plants can be stacked on top of each other, resulting in greater food supply. Figure 10b shows crop irrigation. This is where plants are watered, enabling them to grow. This is done in arid countries such as Burkina Faso and Egypt. In Egypt, water is pumped out of the River Nile and used to irrigate crops, resulting in greater crop yields and a larger food supply. Drip irrigation is used in some countries; this is where plants are watered individually, enabling plant growth but also ensuring water is not lost by evaporation or surface run-off.

05.1 **Any one from:** Spain; Greece; Macedonia **[1]**

05.2 C **[1]**

05.3 Several countries are in northern Europe **[1]**; there is a cluster of countries in south-east Europe **[1]**; most countries are found next to the coast **[1]**

05.4 The climate of a country will affect how much water a country has **[1]**. For example, arid climates such as those in Egypt will limit water supply **[1]**. Many LICs lack the money to invest in water infrastructure [1], and this can limit access to water supplies [1]. Some countries have large populations [1], which means that demand is greater than supply [1]. Rapid industrialisation in countries such as India [1] is leading to higher levels of water demand, putting pressure on supplies [1].

05.5 Not having access to a reliable supply of water [1]; a lack of water [1]; not being able to access clean water [1]

05.6 This question is level marked. Your answer must:
- refer to Figure 12a and 12b
- include a range of well explained ideas.

Example points:
Figure 12a shows a desalination plant. Desalination is where seawater is turned into freshwater, which can then be used for human consumption or irrigation. Salt and other minerals are removed from the seawater. This strategy can be used by coastal countries (such as Spain and Saudi Arabia) to increase their water supply but is expensive. Figure 12b is a reservoir in the UK. Reservoirs can be built in upland areas. Dams are constructed across river valleys, holding water back and creating a reservoir. This is a way of catching more water and therefore increasing water supply. It can also lead to other benefits such as the generation of hydroelectric power. Artificial reservoirs can also be used in lowland areas (for example, Grafham Water in Cambridgeshire). Again, this strategy is expensive.

06.1 India [1]

06.2 B [1]

06.3 They are all in the northern hemisphere / north of the Equator [1]; they are scattered across three different continents [1]

06.4 In some countries, energy costs are expensive [1]; this means people try to use energy more efficiently [1]. In many African countries, the population is growing rapidly [1] and this will lead to greater demand for energy [1]. Countries such as China and India are industrialising [1], so there is greater demand for energy in industry [1]. As the standard of living increases in countries such as China [1], people buy more technological goods, causing an increase in demand for energy [1]. To mitigate climate change [1], countries such as the UK are encouraging people to use less energy [1].

06.5 A lack of energy [1]; not being able to access affordable and reliable energy supplies [1]

06.6 This question is level marked. Your answer must:
- refer to Figure 14a and 14b
- include a range of well explained ideas.

Example points:
One way in which a country can increase its energy supply is by developing renewable sources of energy. The UK is rapidly developing wind energy through the construction of wind farms, both onshore and offshore, as shown in Figure 14a. The UK is Europe's windiest country and therefore by harnessing this energy and turning it into electricity, it will help to boost energy supply. This energy does not create any carbon dioxide emissions and therefore it helps to mitigate climate change. Unfortunately, this type of energy is not reliable, as sometimes there is no wind. France and South Korea have developed tidal power plants, while Iceland's main sources of energy are hydroelectricity and geothermal. Figure 14b shows a nuclear power plant in France. Nuclear power makes up over 70% of France's energy mix. Nuclear energy is created by using uranium. Nuclear fission splits uranium atoms, which generate heat. This heat is used to create steam that drives electric turbines. The UK is building another nuclear power station at Hinkley Point in Somerset and another is planned at Sizewell in Suffolk. Whilst this type of energy is non-renewable, it is very efficient and so will help to increase energy supply.

Pages 213–224
Paper 3 Geographical applications
Section A: Issue evaluation

01.1 A [1]

01.2 Coal, oil and gas [1]; energy sources formed by the remains of plants and animals from millions of years ago [1]

01.3 Just less than 50% of gas is produced in the UK [1]; over half of the UK's gas is imported [1]; imports of gas come from a variety of different countries [1]; most gas is imported from Norway [1]

01.4 Conflicts between two countries [1] could lead to a disruption in supply [1]. The cost of energy is likely to be more expensive [1];

this will lead to the cost of other goods and services increasing as well [1]. Higher energy costs will mean people have less money to buy other goods and services [1].

01.5 This question is level marked. Your answer must:
- cover both economic and environmental issues
- include a range of well explained ideas
- use examples.

Example points:
Economically, the exploitation of energy sources can affect job availability. Most coal mines in the UK have closed. This has led to high unemployment in areas such as south Wales and north-east England. Job losses can lead to a negative multiplier effect, triggering further job cuts and a reduction in tax revenue for local authorities. In contrast, as renewable sources of energy are developed, jobs can be created. Another economic issue is the cost of energy. As supplies of North Sea oil and gas decrease, the UK is importing more energy from countries such as Norway and Qatar. This is leading to rising energy prices. Environmentally, energy exploitation can have negative effects. For example, coal mining and the building of wind turbines can affect wildlife and therefore food chains and ecosystems. Closed mines and open-cast operations can leave waste heaps and pits. However, tips can be reused for ski slopes and nature trails, and pits for sailing and water sports. Exploitation of energy often involves building infrastructure, affecting the appearance of the landscape. For example, many people don't like wind turbines owing to their visual impact. Changing the types of energy we exploit can bring benefits to the environment. Reducing the use of fossil fuels and developing the use of cleaner renewable energy sources will reduce carbon emissions, helping to mitigate against climate change and its negative effects.

02.1 Sources of energy that will last forever [1]; infinite resources [1]; sources which once established don't release any pollutants / carbon dioxide emissions [1]; sources such as solar, wind and geothermal power [1]

02.2 This question is level marked.
Your answer must include a range of well explained ideas.

Example points:
One reason for the UK developing renewable sources of energy is to try to reduce the country's carbon footprint. Once renewable sources are established, they don't release greenhouse gases such as carbon dioxide. This will help the UK to reach targets set by the Paris Agreement and it will also mitigate climate change. Another reason for the UK developing renewable sources of energy is to increase energy security. Supplies of North Sea oil and gas are running out so the UK will need new sources of energy to replace them. Furthermore, developing sources of energy in the UK will reduce the need to import energy from other countries. This will make our energy supply more reliable as the UK becomes less dependent on other countries. In the long run, this should lead to a decrease in energy costs.

03.1 D [1]

03.2 This question is level marked. Your answer must:
- refer to Figure 4
- include several well developed reasons.

Example points:
The UK is an island so it is surrounded by water. This means there are lots of areas offshore where wind turbines can be built. These areas are exposed, providing the ideal conditions for wind energy production. Offshore there is also lots of space, which means that wind turbines don't occupy land that can then be used for other purposes. Figure 4 shows some wind turbines located in upland areas. Again, these areas are exposed, creating good conditions for wind energy generation. The UK has got lots of upland areas, such as Snowdonia in Wales and the Lake District in north-west England.

03.3 This question is level marked. There are also 3 marks for spelling, punctuation and grammar. Your answer must:
- indicate whether you are for or against the statement
- refer to evidence from the resources booklet
- include a range of well explained ideas
- include a conclusion where you summarise your overall answer to the question.

It does not matter whether you agree or disagree with the statement as long as you can provide well explained reasons.

Example points:
I am against the development of wind energy in the Scottish Highlands.
One reason is that wind energy in not reliable as it depends on there being wind. In Resource 3, it mentions that, in January 2017, the 'wind died for a week'. This would result in energy insecurity and could cause problems for people and businesses. Investment in nuclear power would be better as this type of energy is more reliable.
Resources 7 and 8 refer to some of the negative environmental impacts of wind energy generation. The Scottish Highlands is an area of natural beauty. I think that this should be protected. Developing wind turbines would change the appearance of the landscape negatively. This may deter visitors, affecting tourism in the area. This could lead to a loss of trade for businesses relying on tourism, such as hotels, which may lead to job losses. It could also have a negative effect on the price of local people's houses. Building wind turbines may disturb habitats, which could drive wildlife out of the area. Furthermore, as mentioned in Figure 8, the construction of wind turbines can affect bird flightpaths and could even kill some birds. This would affect food chains.
I am not completely against the development of wind energy. It does bring benefits through job creation and, once built, this type of energy does not create carbon emissions and this will help to mitigate climate change. As Resource 6 shows, the amount of electricity being created by wind power is increasing. However, I don't think these turbines should be built in areas of natural beauty. They could be built in areas that have already been developed; for example, near to motorways or industry.
Overall, I am against building wind turbines in the Scottish Highlands. However, I am not against developing wind energy. I think that the UK should develop wind energy, along with other sources such as solar and nuclear. This will create a diverse energy mix, ensuring future energy security.

Section B: Fieldwork

04.1 C [1]

04.2 It might be difficult to get to [1] because the land may be very steep [1]. The land might be private [1] so not accessible to the public [1]. There may be animals in the area [1], so it would be dangerous [1].

04.3 Bar chart [1]
Easy to compare different locations [1]; easy to see differences between different locations [1]; consists of discrete/discontinuous data [1]; gives an accurate number for each site [1]

04.4 Advantage – considers all the data [1]; you do not need to arrange the data as you do when you calculate the median [1] Disadvantage – very small or large values can affect the mean [1]; anomalous values can distort the mean [1]; it may not give a true central value of the data [1]

04.5 There is a positive correlation [1]. As distance down the valley increases, the mean length of stones increases [1]. The mean length of stone increases by nearly three times between Location A and Location E [1].

04.6 Correct completion of the graph (i.e. 'Good' 22 to 40 and 'Very good' 40 to 50). Must use shading as indicated on the key. [1]

04.7 56% [1]

04.8 It is easy to see differences in the opinions of people about the quality of the two footpaths [1]; it provides accurate numbers [1]; it is an appropriate method for discrete/discontinuous data [1].

04.9 This question is level marked. Your answer must:
- refer to Figure 9
- identify similarities and differences between the width of the two footpaths.

Example points:
Figure 9 shows that both footpaths are narrower at the start and at around 900 metres along the footpath. Both footpaths are at their widest at about 700 metres. Footpath A is always wider than Footpath B. Footpath A starts to get wider after 200 metres whilst Footpath B starts to get wider after about 300 metres. Footpath A narrows at 900 metres whereas Footpath B narrows at 800 metres.

05.1 A risk is drowning in a river / sea [1]. This risk can be reduced by checking river levels / tide times [1] and by staying back from the water's edge [1]. A risk is tripping on uneven ground [1]. This risk can be reduced by wearing appropriate footwear [1]

and staying on designated footpaths [1]. A risk is cliff rock fall [1]. This risk can be reduced by staying back from the cliffs [1] and wearing protective headgear [1]. A risk is severe sunburn / sunstroke [1]. This risk can be reduced by wearing a hat [1] and regular application of sun cream [1].

05.2 This question is level marked. Your answer must:
- refer to the method used in a physical geography fieldwork investigation
- refer to specific data collection methods that you used
- recognise the advantages and disadvantages of your data collection method(s)
- include a range of well explained ideas.

Example points:
One method of data collection I used was river velocity measurements. I carried out these measurements at three systematically-spaced locations along the River X in Cheshire. I used a flow meter; this should mean the measurements are accurate. At each location I measured the velocity three times and then calculated the mean. This reduced the influence of anomalies, making the data more reliable. I made sure that I measured the speed of the river where I thought it was fastest flowing, ensuring a consistent approach at all three sites. By collecting data at three sites downstream, I was able to answer my hypothesis that river velocity increased downstream. A weakness of my data collection method was that, by only doing three readings, an anomalous reading could have a significant impact on my data. It was hard to identify where the fastest flowing water was occurring, so I may not have carried out my measurements in the correct location. Furthermore, three sampling sites is a small sample size so my data may not be a true representation of the river. The data was also only collected at one time in the year after several weeks of little rain, which may not be representative of other times.

05.3 **Example points:** It was within walking distance of the school [1], so it did not take long to get there [1]. There were no busy roads nearby [1], so it was a safe area in which to collect data [1]. There was a range of survey points available [1] with enough variation that we could look at how environmental quality changes over distance [1].

05.4 This question is level marked. There are also 3 marks for spelling, punctuation and grammar. Your answer must:
- refer to either your physical or human geography fieldwork investigation
- evaluate the degree to which the data collected allowed you to reach valid conclusions
- refer to specific evidence
- include a range of well explained ideas
- include a conclusion where you summarise your overall answer to the question.

Example points:
The title of the fieldwork investigation was 'Does quality of life increase with distance from the town centre?'.
I collected housing and environmental quality data at 10 systematically-spaced locations on a transect going through the town of X. The sample size was large enough that I could present the data in the form of scatter graphs. These clearly showed a positive correlation, meaning that environmental quality and housing quality increased with distance from the town centre. There were some anomalies, so I could not completely agree with my hypothesis.
Whilst the data I collected enabled me to reach conclusions, there are limitations with my data which may mean that these conclusions are not completely valid. One limitation of my study is that I only studied one transect within the town. Other areas may not demonstrate the same characteristics, meaning that my data may not be representative of the whole town. Many of the surveys I carried out, such as the environmental quality survey, were subjective and just based on my opinion. This means that the data may be biased and therefore the conclusions I have drawn may not be entirely valid. Furthermore, some of the secondary data I used, census data, was from 2011 and therefore out of date. It may no longer represent the characteristics of the people that live in the area today.
In conclusion, the data I collected did enable me to reach conclusions, although they may not be entirely accurate.

Acknowledgements

The authors and publisher are grateful to the copyright holders for permission to use quoted materials and images.
Page 5 (map): © Crown copyright and database rights (2022) Ordnance Survey (AC0000808974)
Page 14 (image): GOES 12 Satellite, NASA, NOAA
Page 15 (image): Nilfanion [Public domain], courtesy of Wikimedia Commons
Page 16 (image): SSEC/CIMSS, University of Wisconsin–Madison (courtesy of Wikimedia Commons)
Page 18: © Crown copyright. Contains public sector information licensed under the Open Government Licence v3.0
Pages 47, 50, 51, 52, 64, 65: courtesy of Janet Hutson
Page 73: © Paul Gillett, geograph.org.uk/p/3119879
Page 85: Robert Walker / Adnams Brewery Distribution Centre opened 2006 / CC BY-SA 2.0
Page 96: © fotoshoot / Alamy Stock Photo
Pages 108, 109, 120: courtesy of Brendan Conway
Page 150: © Citynoise at English Wikipedia/wikipedia.org
Page 156: © Alan Curtis / Alamy Stock Photo
Page 171: © Aflo Co. Ltd. / Alamy Stock Photo
Page 175: © Nick Hanna / Alamy Stock Photo; © ZUMA Press, Inc. / Alamy Stock Photo; © John Peter Photography / Alamy Stock Photo
Page 180: © Paralaxis / Alamy Stock Photo
Page 182: © Crown copyright and database rights (2022) Ordnance Survey (AC0000808974)
Page 184: © David Thompson / Alamy Stock Photo; geogphotos / Alamy Stock Photo
Page 185: © Crown copyright and database rights (2022) Ordnance Survey (AC0000808974)
Page 187: © Ian Goodrick / Alamy Stock Photo; © Paul Glendell / Alamy Stock Photo
Page 188: © Crown copyright and database rights (2022) Ordnance Survey (AC0000808974)
Page 190: © Neil McAllister / Alamy Stock Photo; © Vivienne Crow / Alamy Stock Photo
Page 194: © Paul Lovelace / Alamy Stock Photo
Page 195: © Dave Ellison / Alamy Stock Photo; © Mark Waugh / Alamy Stock Photo
Page 206: © Orapin Joyphuem / Alamy Stock Photo; © Erberto Zani / Alamy Stock Photo
Page 209: © Shine-a-light / Alamy Stock Photo; © Andrew Woodacre / Alamy Stock Photo
Page 212: © aerial-photos.com / Alamy Stock Photo; © Paul Glendell / Alamy Stock Photo
Page 217, 226 © Bjmullan/commons.wikimedia.org
Page 225: US Energy Information Administration, based on Digest of UK Energy Statistics and National Statistics: Energy Trends
Page 230: Map: Crown copyright 2021 / licensed under terms of the Open Government Licence

All other images Shutterstock.com and HarperCollinsPublishers

Every effort has been made to trace copyright holders and obtain their permission for the use of copyright material. The authors and publisher will gladly receive information enabling them to rectify any error or omission in subsequent editions. All facts are correct at time of going to press.

Published by Collins
An imprint of HarperCollinsPublishers Ltd
1 London Bridge Street
London SE1 9GF

HarperCollinsPublishers
Macken House, 39/40 Mayor Street Upper, Dublin 1, D01 C9W8, Ireland

© HarperCollinsPublishers Limited 2022

ISBN 9780008535001

First published 2022

10 9 8 7 6 5 4 3 2

Publisher: Clare Souza
Authors: Paul Berry, Brendan Conway, Janet Hutson, Dan Major, Tony Grundy, Robert Morris and Iain Palôt
Videos: Eleanor Barker
Project Management and Editorial: Richard Toms
Cover Design: Sarah Duxbury and Kevin Robbins
Inside Concept Design: Sarah Duxbury and Paul Oates
Text Design and Layout: Ian Wrigley and Jouve India Private Limited
Artwork: Ian Wrigley, Sarah Duxbury and Jouve India Private Limited
Production: Lyndsey Rogers
Printed in the UK, by Martins The Printers

MIX
Paper | Supporting responsible forestry
FSC™ C007454

This book is produced from independently certified FSC™ paper to ensure responsible forest management.

For more information visit:
www.harpercollins.co.uk/green